BURLEIGH DODDS SCIENCE: INSTANT INSIGHTS

NUMBER 39

Crops as livestock feed

I0130690

burleigh dodds
SCIENCE PUBLISHING

Published by Burleigh Dodds Science Publishing Limited
82 High Street, Sawston, Cambridge CB22 3HJ, UK
www.bdspublishing.com

Burleigh Dodds Science Publishing, 1518 Walnut Street, Suite 900, Philadelphia, PA 19102-3406, USA

First published 2021 by Burleigh Dodds Science Publishing Limited
© Burleigh Dodds Science Publishing, 2021. All rights reserved.

Notice
No responsibility is assumed by the publisher for any injury and/or damage to persons or property as a matter of product liability, negligence or otherwise, or from any use or operation of any methods, products, instructions or ideas contained in the material herein.

British Library Cataloguing in Publication Data
A catalogue record for this book is available from the British Library

ISBN 978-1-80146-171-9 (Print)
ISBN 978-1-80146-172-6 (ePub)

DOI 10.19103/9781801461726

Typeset by Deanta Global Publishing Services, Dublin, Ireland

Contents

Series list

Title	Series number
Sweetpotato	01
Fusarium in cereals	02
Vertical farming in horticulture	03
Nutraceuticals in fruit and vegetables	04
Climate change, insect pests and invasive species	05
Metabolic disorders in dairy cattle	06
Mastitis in dairy cattle	07
Heat stress in dairy cattle	08
African swine fever	09
Pesticide residues in agriculture	10
Fruit losses and waste	11
Improving crop nutrient use efficiency	12
Antibiotics in poultry production	13
Bone health in poultry	14
Feather-pecking in poultry	15
Environmental impact of livestock production	16
Pre- and probiotics in pig nutrition	17
Improving piglet welfare	18
Crop biofortification	19
Crop rotations	20
Cover crops	21
Plant growth-promoting rhizobacteria	22
Arbuscular mycorrhizal fungi	23
Nematode pests in agriculture	24
Drought-resistant crops	25
Advances in crop disease detection and decision support systems	26
Mycotoxin detection and control	27
Mite pests in agriculture	28
Supporting cereal production in sub-Saharan Africa	29
Lameness in dairy cattle	30
Infertility/reproductive disorders in dairy cattle	31
Antibiotics in pig production	32
Integrated crop–livestock systems	33
Genetic modification of crops	34

The use of feedlot/cereal grains in improving feed efficiency and reducing by-products such as methane in ruminants

Kristin Hales, US Meat Animal Research Center – USDA-ARS, USA; Jeferson Lourenco, Darren S. Seidel, Osman Yasir Koyun, Dylan Davis and Christina Welch, University of Georgia, USA; James E. Wells, US Meat Animal Research Center – USDA-ARS, USA; and Todd R. Callaway, University of Georgia, USA

1 Introduction

Cattle are a marvel of evolutionary processes because they convert low-quality feedstuffs, such as forages, into high-quality protein sources for humans. The ability of the ruminant to convert sunlight into meat, milk, and fiber is extraordinary and is mediated through the symbiotic relationship between the host animal and its resident gastrointestinal microbial consortium (Hungate, 1966). The stepwise degradation of feeds by the members of this mixed microbial microorganism fermentation is crucial to production of volatile fatty acids (VFAs) upon which the animal depends for energy (Russell, 2002). However, the molar proportions of each VFA produced by the ruminal microbial ecosystem vary depending on which dietary ingredient or feedstuff is fermented by the resident microbial population,

http://dx.doi.org/10.19103/AS.2020.0067.23

which can have profound impacts on animal production efficiency, carcass quality, and food safety (Russell and Hespell, 1981; Depenbusch et al., 2008; Verdu et al., 2015; Wilson et al., 2016). While the ruminant animal clearly evolved to degrade forages, the ability to ferment feeds is not limited to forages, but includes the ability to degrade cereal grains such as wheat, corn, barley, sorghum, and oats.

The environment of the rumen is highly reduced (Russell, 2002), meaning that reducing equivalent disposal (e.g. NADH NAD) can become limiting, imperiling the continuation of the fermentation of feedstuffs. Interspecies hydrogen transfer to regenerate NAD results in methane production (Iannotti et al., 1973; Thiele and Zeikus, 1988), and is a clear keystone to the symbiotic relationships that occur within the ruminal microbial population. However, the fundamental importance of interspecies hydrogen transfer to methane underlines the critical role of methane as a reducing equivalent sink to allow anaerobic degradation of feedstuffs to continue unless an alternative electron sink is provided. Despite the benefits the ruminal microbial population derives from methane production, it is both a significant loss of carbon and energy, representing from 2% to 8% of the GE to the animal (Dong et al., 2006) and in some situations methane can represent up to 12% of the DE (Johnson and Johnson, 1995), especially when cattle are fed forages. It should be noted that methane is a potent and notable greenhouse gas that is of increasing global importance.

The contribution of livestock to global climate change is caused by either the direct (e.g. enteric fermentation) or indirect (e.g. feedstuff production) emission of greenhouse gases (GHGs) (Steinfeld et al., 2006; Beauchemin et al., 2009; Haque, 2018). Carbon dioxide (CO_2), methane (CH_4), and nitrous oxide (N_2O) are the main GHGs produced by the livestock over the course of their production (Hungate, 1966). Among livestock, ruminants are the key contributors to the GHG, approximately 80% of the total emissions in the sector (Opio et al., 2012), while pigs and poultry only have approximately 9% and 8% contribution, respectively (Gerber et al., 2013). Ruminants consume plants that utilize CO_2 in photosynthesis; therefore, the CO_2 emitted by animals is not viewed as a net contributor to the climate change (Steinfeld et al., 2006). Consequently, CH_4 and N_2O are considered the primary GHGs caused by enteric fermentation and feed production in ruminant production systems, having massive global warming potentials (GWP) of 25 and 298 CO_2 equivalent, respectively (Forster et al., 2007; Eckard et al., 2010).

Since the mid-1940s, the US beef production industry has increasingly relied upon the use of grain feeding to produce a wholesome, high-quality protein source. By feeding starch-containing feeds such as corn, the end products of the microbial fermentation are altered, with significant impacts upon fermentation efficiency and host physiology. Starch fermentation yields a greater proportion of propionate, which leads to increased levels of intramuscular marbling in grain-fed cattle. Other chapters of this book will address the end products of the rumen microbial fermentation and inefficiencies inherent to the ruminal fermentation, but in this chapter, we will address the impacts of feeding grain to cattle to reduce deleterious fermentation by-products such as methane.

2 Types of cereal grains fed to cattle

Beef production in the United States is localized largely in the central one-third of the nation, primarily in the Great Plains. The proximity to the Grain Belt means that shipping grain to the cattle is logistically and economically feasible, therefore many of our beef cattle are grown in feedlots located in the central high plains. Cereal grains are a large part of the US agricultural economy with over 475 million metric tons produced in 2016, including wheat, corn, barley, sorghum, rice, and oats. While some cereal grain production is for human consumption, a large portion is used in the livestock feed industry because of the ability to meet high-energy demands associated with modern levels of production. Starch is often the primary nutrient used to promote increased levels of production in ruminant diets, and cereal grains are an effective source of starch (Theurer, 1986).

The kernel of cereal grains primarily consists of three main parts including the bran (or outer hull), the endosperm (primary location of starch), and the germ (or embryo). Although grain is a good source of starch, the cell wall (outer layer) is virtually indigestible in livestock and often requires further processing before feeding. Some common processing procedures include dry rolling, grinding, cracking, steam rolling, steam flaking, pelleting, and the addition of enzymes. Dry rolling, cracking, and grinding are often grouped together because each of these mechanical processes increases the surface area of the grain for digestion by breaking the seed coat and reducing particle size (Rowe et al., 1999). Steam rolling includes heating the grain through a steam cabinet to soften the kernels before rolling. Steam flaking involves processing the corn under steam pressure before rolling into a flake. This latter method partially gelatinizes the present starch as well as breaks the seed coat and endosperm, making the grain more susceptible to degradation by amylolytic bacteria (Rowe et al., 1999). Pelleting combines smaller particles together to make a large particle that allows control of site and rate of digestion by altering the density and size of the final particle created (Rowe et al., 1999). Among the list of grains produced in the United States each year, the primary cereal grains used in livestock rations are barley, corn, and wheat.

2.1 Barley

Barley (*Hordeum vulgare L.*) ranks fourth in the production quantity after corn, rice, and wheat. In 2014, the production of barley totaled over 144 million tons with the top five producers being Russia, France, Germany, Australia, and Ukraine. Although total acres of barley harvested has steadily decreased in the last five years, more productive varieties and farming practices have allowed the total quantity to increase. Barley's nutritional value makes it an appropriate supplement to cattle, especially in systems where high levels of production are optimal. Starch generally makes up between 52% and 60% of the barley kernel while protein and energy are around 13% and 85%, respectively. Although cereal grains are typically low in calcium and high in phosphorus, barley provides a greater amount of calcium and ß-glucans when compared to corn, wheat, and sorghum. Inside the barley grain, multiple types of starch have been identified including waxy, normal, and high amylose (Zhu, 2017). Although the nutritional value of barley is beneficial for feeding cattle, the kernel is surrounded by a fibrous hull that is rather indigestible in the rumen, requiring some processing, such as dry rolling, to maximize utilization (Beauchemin et al., 1994). Some studies have shown that dry rolling barley can increase digestibility from whole barley by 32.7% (Toland, 1976).

2.2 Corn

Corn (*Zea Mays L.*) is the most common cereal grain used in livestock feed because of the nutritional value, productivity, and availability. Not only can corn grain be fed to cattle (processed or whole), but there are multiple forms and by-products available from corn production including distiller's grain (wet and dry), corn silage, and corn gluten (Firkins et al., 1985; Fron et al., 1996). The livestock feed industry benefits greatly from many corn by-products that are created during the production of ethanol such as stillage, condensed distiller's solubles, wet distiller's grain, and dry distiller's grains. Stillage is a liquid product that is created from mash distillation that can be used to replace water in cattle diets to decrease dry matter intake without compromising performance. Stillage can also be dehydrated to create condensed distillers solubles or syrup. The solid fraction created from ethanol production is known the distillers grain. Distiller's grain is a highly digestible alternative to whole corn and has a higher energy value because of the increased fat content. The mechanical processing of corn increases digestibility by 5–10% and increases nutrient availability by cracking the pericarp (smooth outer surface), that is basically indigestible in the rumen, to allow microbes to access the starch inside the kernel. While whole corn generally consists of 70–72% starch, 10% protein, and 87% TDN, the crude protein content of corn by-products can range from 20% to 35% and TDN from 70% to 100% (Beauchemin et al., 1994).

2.3 Wheat

Wheat (*Triticum aestivum L.*) is an important part of the livestock feeding industry because of its versatility. While most cereal grains are planted in the spring and grow during the summer, wheat can be planted in the fall to provide a good source of grazing for livestock during the winter. Aside from grazing, wheat can be used to make hay, silage, or can be fed as a grain. The wheat seed is generally made up of 60% starch and has approximately 16% protein and 80% TDN or energy. The wheat kernel is composed of three parts: bran (outer), endosperm (85–86% starch), and the germ (embryo). The bran, or outer layer of the wheat kernel, can escape ruminal fermentation so further processing, such as steam flaking or dry rolling, is required (Kreikemeier et al., 1990). If whole wheat is fed, chewing time tends to increase as well as dry matter disappearance or digestibility tends to decrease. Much like other cereal grains, wheat can be used as an effective source of energy in rations but calcium supplementation may be required since wheat is low in calcium and high in phosphorus. Compared to other grains, wheat has shown, when further processed, to have the greatest dry matter disappearance in animal studies (Herrera-Saldana et al., 1990).

2.4 Other cereal grains

Oats (*Avena sativa*) rank sixth in global cereal grain production after wheat, corn, rice, barley, and sorghum (Stevens et al., 2004). Much like corn, oats have many different uses in the livestock feeding industry including grazing, fodder, silage, haylage, straw, and hay. Although there are other uses for oats, Stevens et al. (2004) reported that 74% of oats are used in livestock feed. Similar to barley, oats offer a valuable source of ß-glucans and mineral, while providing around 13% crude protein. Although oats have a softer kernel than other cereal grains, which creates difficulty in the milling process, the kernel is composed of 60% starch which makes it an effective source of energy in livestock feeds. Sorghum

(*Sorghum bicolor*) is not used as extensively as other cereal grains (Etuk et al., 2012). Sorghum offers a higher percentage of starch than oats (74% vs. 60%) with slightly less protein (12.3% vs. 13%). While sorghum behaves in a similar fashion to corn when fed to livestock, when fed as a forage, grain, or silage, only 51% of the sorghum produced is used as livestock feed. Consideration of plant maturity when feeding sorghum is critical as young, green plants contain a compound called dhurrin. Dhurrin is a cyanogenic glycoside that yields hydrogen cyanide (HCN) during hydrolysis which is lethal to animals (Etuk et al., 2012).

3 Cereal grain production

Cereal grain production is a vital part of the world economy with over 600 million ha of production in 2010. However, all cereal grains are not produced in every part of the world as there are several factors that determine where crops are produced including environmental, cultural, and economic (Awika, 2011). The main factor that determines crop production location is the local environmental conditions.

3.1 Corn

Corn is the most important of all cereal grains in terms of production with over 800 million metric tons produced in 2010 with 40% produced by the United States (Awika, 2011). Corn is relatively drought and frost susceptible, so it is grown in areas with plenty of rainfall or ability to irrigate and is planted in the spring which allows it to grow in the warm summer months. Corn is produced in a wide variety of states in the United States with the highest concentration being in the Midwest states or the 'corn belt'. The states that make up the 'corn belt' are primarily Indiana, Illinois, Iowa, Minnesota, Missouri, Nebraska, and Kansas with some surrounding states being included as well.

The United States Department of Agriculture (USDA) defines a value-added product as changing the physical state or form of a product to enhance its value. In this sense, cattle serve as a tremendous value-added product of corn. Cattle have evolved to form a symbiotic relationship with the microbes that inhabit their rumen and allow them to efficiently survive on feedstuffs that would normally be unavailable to the animal. The microbes inside the rumen can degrade the starch present in corn (and all cereal grains) to produce VFA that provide energy to the animal and promote growth at a conversion rate of 5–7 pounds of feed per pound of gain, roughly. Feed corn commonly grown is relatively unsuitable for human consumption, but feeding to cattle changes the form of that product into a highly digestible animal protein and adds significant value. For example, the average bushel of corn weighs 56 pounds and is currently worth $3.61 per bushel, which is roughly $0.065/lb. The current live cattle prices are $1.26/lb while a grain-fed, choice ribeye steak is worth $9.99/lb from a popular grocery store in the United States. With this information, it can be calculated that 1000 pounds of corn is worth $65, and 1000 pounds of live cattle are worth $1260, though this vastly simplified calculation ignores some important other costs (e.g. varying feed conversion efficiency, cost of maintaining feedlots, transport, opportunity cost, debt service, etc.) that are associated with producing cattle to market weight and bringing the product to consumers' plates. Corn is an important part of cattle rations because of its ability to add significant amounts of energy in the production systems where high levels of performance are demanded. With the vast difference in price for corn and cattle, producers could certainly benefit from viewing cattle as a value-added product of corn.

3.2 Wheat

Wheat is arguably the most versatile cereal grain. Wheat can be grown in a wider range of environments as it is more resistant to drought and temperature variation so it can be grown in both winter and summer months (Awika, 2011). Because of its versatility across different climates and temperatures, wheat production is distributed across parts of the United States and Canada. The 'wheat belt' extends from central Texas to central Alberta including the states of Oklahoma, Kansas, Nebraska, Montana, and the Dakotas.

3.3 Barley

Barley is more temperature tolerant and can survive in colder climates than other cereal grains. The majority of barley production in the United States is in the north/northwestern states of Montana, North Dakota, Idaho, and Washington with an average of 205 million bushels produced between 2008 and 2012. In North America, Canada produces the majority of barley as they produce about 40% more than the United States.

3.4 Other cereal grains

Sorghum and oats are produced in much lower quantities in the United States when compared to corn, wheat, barley, and even rice (although rice is not generally used as a livestock feed). The majority of oat production in North America, much like barley, occurs in Canada. The highest concentration of oat production in the United States takes place in Iowa, Minnesota, South Dakota, North Dakota, and Wisconsin. Sorghum is rather tolerant to heat and drought when compared to other cereal grains which makes it a popular crop in African countries (Awika, 2011), and in the United States it is mainly grown in dry lands of the 'sorghum belt' that stretches from southern Texas to South Dakota including Oklahoma, New Mexico, Colorado, Kansas, and Nebraska.

4 Dietary factors affecting methane production by ruminants

Most of the enteric CH_4 is generated by ruminants via ruminal fermentation of carbohydrates, proteins, and, to some extent, lipids by the bacterial, protozoal, and fungal populations under anaerobic conditions, leading to the production of VFAs, mainly acetate, propionate, and butyrate that are used by the animal as a source of energy. The production of gases (CO_2, H_2, and CH_4) is a by-product of fermentation, which is removed through eructation (Boadi et al., 2004; Kebreab et al., 2006; Martin et al., 2010). Fermentation is an oxidative process, during which reduced cofactors (NADH, NADPH, FADH) are re-oxidized (NAD^+, $NADP^+$, FAD^+) through dehydrogenation reactions generating H_2 in the rumen. Once produced, H_2 is utilized by methanogenic archaea, a microbial group distinct from Eubacteria, to reduce CO_2 or formate to CH_4 (McAllister and Newbold, 2008; Martin et al., 2010). However there are some ruminal methanogens such as Methanomassiliicoccaceae that generate methane by reducing methyl groups from other metabolites (e.g. methylamines) (Martinez-Fernandez et al., 2018). This route can be important to ruminal methane production because pectin contains a higher ratio of methylated compounds. Moreover, production of acetate liberates H_2, whereas propionate serves as a net H_2 sink. Consequently, diets that increase propionate and decrease acetate in the rumen are often associated with a reduction in ruminal CH_4 production, given that less H_2 is available to methanogens for reducing from CO_2 to CH_4 (Beauchemin et al., 2009).

The amount of emitted CH_4 by the ruminants depends on various factors including carbohydrate intake, type of the carbohydrate sources, ruminal pH, residence time in the rumen, rate of ruminal fermentation, and rate of methanogenesis. The daily emission of enteric CH_4 can be inhibited by reducing feed intake and/or fermentation rate in the rumen; however, animal growth performance should be considered. Fat supplemented diets lower the carbohydrate fermentation in the rumen due to the replacement of carbohydrate with lipids; however, fiber digestibility may be adversely affected, leading to decreased feed conversion efficiency in the ruminant (Martin et al., 2008; Beauchemin et al., 2009). Dry matter intake (DMI) is another major factor affecting methane production in ruminants, and a positive relationship of DMI and methane production is reported. However, methane production per unit of intake (g CH_4 /kg DMI) decreases with increasing DMI, suggesting a higher rumen turnover leading to a lower digestibility of the diet (Buddle et al., 2011).

Type of carbohydrate plays a critical role in CH_4 production as it can impact ruminal pH and subsequently alter the microbiota present. The digestibility of cellulose and hemicellulose is highly related to methane production compared to soluble carbohydrate (Hook et al., 2010). A positive relationship was reported between digestibility of hemicellulose and methane emission in non-lactating cows fed forage; however, a negative relationship was reported between digestibility of cellulose and methane emission (Holter and Young, 1992). Another study concluded that there seems to be a curvilinear relationship between methane production and proportion of concentrate in the ration, with methane losses of 6–7% of gross energy (GE) remaining constant at 30–40% concentrate levels in the ration and then reducing to 2–3% of GE with a concentrate level of 80–90% (Sauvant and Giger-Reverdin, 2007). On the other hand, in dairy cows increasing the level of concentrates to inhibit CH_4 emitted is not promising due to the fact that milk quality is adversely affected by the concentrate levels over 50% of the ration (Beauchemin et al., 2008).

It is critical to consider that elevated amounts of rapidly fermentable carbohydrates in a ration can lead to a higher passage rate from the rumen and lower ruminal retention time and pH, directing methanogenesis from slower degrading carbohydrates toward the hindgut and manure (Hindrichsen et al., 2006; Hook et al., 2010). Grinding forage feed prior to the consumption by the ruminants seems to reduce methane production, most likely by increasing the ruminal digestion and flow rate through the gastrointestinal tract (GIT); therefore, less time is available for ruminal methane production (Johnson and Johnson, 1995). Moreover, the ruminal fermentation of rapidly fermentable carbohydrates can improve the production of VFAs; however, if VFA production is higher than the absorption capacity from the rumen, the pH will drop, resulting in a risk of subacute ruminal acidosis (SARA) and imbalance of the rumen microbiota (Plaizier et al., 2008).

5 The role of starch and forage in methane formation

Starch is the primary nutrient used in cattle rations to add increased levels of digestible energy when performance at high levels is required (Theurer, 1986). There are two components of starch, amylose and amylopectin. Amylose is a linear molecule that only accounts for roughly 20–30% of starch and is made up of primarily 1–4 linkages with very few 1–6 branch points (Fig. 1).

Amylopectin, on the other hand, is a larger (although less dense) molecule that is described as 'fluffy' because of its considerable amount of 1–6 linkages in combination with 1–4 bonds as well. Generally, cereal grains are composed of greater than 50% starch with varying sizes of starch granules depending on the ratio of amylose to amylopectin. The size of the starch granule is negatively correlated to the amount of amylose present. Amylose is broken down into sugars by enzymes such as amylases (and ß) and is rapidly fermented to produce VFA (acetate and propionate) and lactate. Amylopectin is also broken down by enzymes such as amylase (and ß), but because of the increased amount of 1–6 branch points, both sugars and limit dextrins are produced (Fig. 2).

The sugars are rapidly fermented inside the rumen while limit dextrans require additional enzymes, such as limit dextranase, for complete degradation. Starch is fermented in the rumen by amylolytic bacteria (i.e. *Streptococcus bovis, Prevotella ruminicola, Ruminobacter* (formerly *Bacteroides) amylophilus, Selenomonas ruminantium, Succinomonas amylolytica*) to produce acetate, propionate, and lactate. As starch is fermented inside the rumen, propionate production increases and the acetate to propionate ratio decreases, which provides more available energy to the animal. While lactate can be detrimental to the animal, due to acidosis, balancing the ration with sufficient forage can prevent many of the negative effects.

Starch is a key source of glucogenic energy for high-yield dairy cows and a vital fuel for rumen microbes in the form of fermentable energy (Koenig et al., 2003). Once in the rumen, starch is mainly broken down by amylolytic bacteria and also by protozoa and fungi to some degree. The enzymes produced by rumen microorganisms are capable of hydrolyzing amylose and amylopectin glycosidic bonds, releasing various glucooligosaccharides. The post-ruminal process of starch breakdown is triggered by pancreatic -amylase secretion, hydrolyzing amylose and amylopectin into dextrins and smaller glucooligosaccharides. The

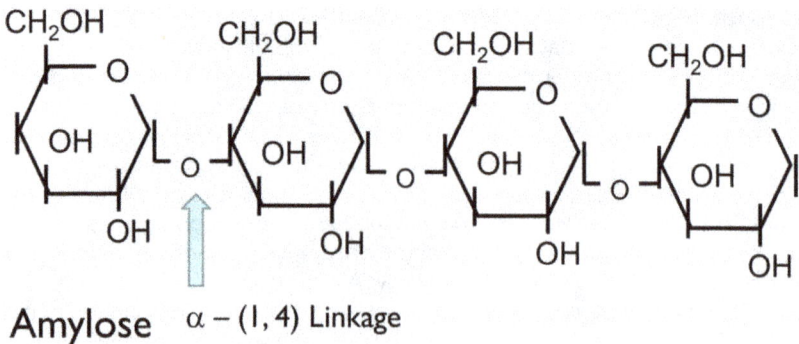

Figure 1 Structure of amylose (starch), characterized by -(1, 4) linkages.

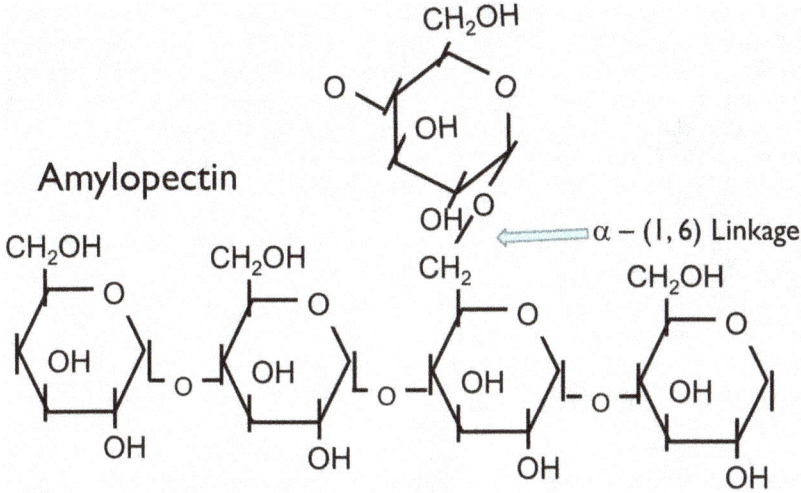

Figure 2 Structure of amylopectin, characterized by -(1, 4) and -(1, 6) linkages.

process is ended by the action of maltase and isomaltase secreted in the intestine, though it is inferred that starch digestion in the ruminant small intestine is limited because evolutionarily little starch reached the small intestine of ruminants (Huntington, 1997; Ortega-Cerrilla and Mendoza-Martínez, 2003; Gómez et al., 2016). The site of starch breakdown impacts the substrates absorbed in ruminants. Ruminal breakdown of starch decreases enteric CH_4 formation, leading to an alternative H_2 sink to methanogenesis, and produces VFAs for absorption and supplies energy to microbial protein synthesis; however, decreased starch digestibility is considered beneficial regarding the preventive effect against acidosis and increase in the supply of glucogenic substrates. On the other hand, starch breakdown in the ruminant small intestine leads to higher energetic efficiency compared to degradation via microbial fermentation due to a reduction in CH_4 production, fermentation heat losses, and greater efficiency of metabolizable energy utilization (Huhtanen and Sveinbjörnsson, 2006; Gómez et al., 2016). It is important to note that lower ruminal starch digestion is not correlated with an increase in its small intestinal digestion; however, it is linked to higher hindgut and lower total tract digestibility (Larsen et al., 2009).

Grain selection in the diet is also critical to the production levels of methane, as well as production efficiency of milk and meat, which offers potential to reduce methanogenesis via current technology and feeding systems (Martin et al., 2010; Moate et al., 2016). Changing the diet of Australian cattle from corn to wheat resulted in a significant decrease in methane yield (g CH_4/kg DMI) (Moate et al., 2018). Researchers have demonstrated that diets containing wheat rather than corn had lower methane yield (11.1 vs. 19.5 g CH_4/ kg DMI, respectively) and methane efficiency (7.6 vs. 15.7 g CH_4/kg milk) (Moate et al., 2019). However, it was noted that some cattle were 'adaptive' to changing the diet over time in regard to milk and methane production metrics, but not all cattle responded the same way (Moate et al., 2018). Thus is apparent that the availability of starch in grains does play an important role in ruminal methanogenesis, but the mode of action of the reduction of methane production is unclear currently, and this offers a distinct opportunity for analysis of the changes in the microbiome and how this can impact methane production.

Forage quality has an impact on CH_4 formation in the rumen. High-quality forage (i.e. young plants) can decrease CH_4 production by changing the fermentation pathway since it has greater levels of readily fermentable carbohydrates and less neutral detergent fiber (NDF), resulting in a higher digestibility and passage rate (Beever et al., 1986). In contrast, mature forage leads to a higher CH_4 production yield (e.g. methane produced/kg DMI) due to a higher C:N ratio, resulting in a decrease in the digestibility (Milich, 1999), which contrasts with the generally accepted relationship that increasing N available with mature forages can increase forage digestibility.

Different types of forage can also affect CH_4 emission because of the differences in their chemical composition. Cereal forages can have substantial levels of starch, favoring propionate production over acetate and can decrease ruminal CH_4 formation (Beauchemin et al., 2008). Legume forage has a lower CH_4 yield due to the presence of condensed tannins, a low-fiber content, a high DMI, and a fast passage rate (Beauchemin et al., 2008); however, results from other studies focusing on the effect of forage type on methane yield are inconsistent (Benchaar et al., 2001; Hammond et al., 2013). It has been stated that C4 grasses produce more CH_4 than the C3 plants (Archimède et al., 2011). Additionally, forage processing and preservation can impact CH_4 emission as well. For example, when forages are chopped or pelleted, they can decrease the CH_4 production per kg of DMI since smaller particles undergo less breakdown in the rumen (Boadi et al., 2004; Martin et al., 2010). Ensiled forages have a tendency to cause less CH_4 formation due to the fact that the digestible oligosaccharides in ensiled forages are fermented during the process (Boadi et al., 2004).

6 H_2 sinks in the rumen and methane production

Methanogenesis is vital for a desirable performance of the rumen because it prevents H_2 accumulation by serving as a reducing equivalent sink. The elimination of methane production would result in an inhibition of dehydrogenase activity involved in the oxidation of reduced cofactors (Wolin, 1975; McAllister and Newbold, 2008). The microbial fermentation of substrates produces various end products that are not equivalent in terms of H_2 output; therefore, accumulated H_2 inhibits microbes in the rumen to oxidize the cofactors having a role in electron transfer in the rumen, resulting in less energy from the fermentation process (Beauchemin et al., 2009; Martin et al., 2010). Acetate and butyrate production in the rumen leads to a release of H_2 and favors CH_4 production, while the propionate formation is a competitive pathway for H_2 utilization (Boadi et al., 2004). Efficiency is a complex production trait to define, but animals that were defined as efficient in gain did not differ in specific methane production or methanogenic bacterial populations from animals that were less efficient in gain (Freetly et al., 2015).

The rumen is an anaerobic fermentation chamber where microbial populations are in a symbiotic relationship, exchanging metabolites to improve each other's growth; therefore, this interaction is called 'cross-feeding' (Schultz and Breznak, 1979). Methane formation is considered as cross-feeding between H_2-producing microbial populations (e.g. fibrolytic fungi and bacteria) and H_2-utilizing methanogens, allowing removal of H_2 and improvement in fiber fermentation (Kobayashi, 2010). Current strategies to reduce methane production need to consider alternative H_2 sinks to methanogenesis. There are several H_2-consuming pathways that occur in the rumen. Methanogenesis is the major one followed by propionate production (i.e. fumarate reduction). Other pathways such as nitrate- and nitrite-reduction, reductive acetogenesis, and biohydrogenation of unsaturated fatty acids have a minor role in H_2 consumption in the rumen (Kobayashi, 2010).

Monensin has gained an interest as a mitigation strategy for CH_4 production, because it has an inhibitory effect on protozoa and gram-positive bacteria, including ruminococci, streptococci, and lactobacilli that are supplying methanogens with substrate for methanogenesis (Russell and Strobel, 1989). The monensin selects for gram-negative microorganisms, leading to a shift toward propionate production in the rumen, therefore, it is hypothesized that monensin does not affect CH_4 production by limiting methanogens (Martin et al., 2010), but inhibits the growth of the bacteria and protozoa supplying a substrate for methanogenesis (Bergen and Bates, 1984; Russell and Strobel, 1989). This hypothesis is supported by the fact that when rumen fluid was administered with monensin in vitro, CH_4 production decreased until H_2 was supplied, at which time CH_4 formation resumed (Russell and Strobel, 1989).

Monensin is generally supplemented to the diet as a premix or is provided via a slow-release capsule (Beauchemin et al., 2008). Studies indicated that the effect of monensin on decreasing CH_4 production seems to be dose-dependent, with lower doses (10–15 ppm) leading to an effective milk response in dairy cows but no effect on CH_4, whereas higher doses (24–35 ppm) decreased CH_4 production (Eckard et al., 2010). However, there have been contradictory results from different studies in terms of the inhibitory effect of monensin in terms of sustainability. In a study, it has been reported that long-term administration of monensin to dairy cows continuously decreased methane by 7% for 6 months without adverse effect on milk yield (Odongo et al., 2007; Ellis et al., 2008). It has been reported that monensin (33 mg/kg) decreased CH_4 emissions in beef cattle by up to 30%, but levels returned to pre-treatment levels within 2 months (Guan et al., 2006).

Dicarboxylic acids such as fumarate and malate are precursors to produce propionate in the rumen and can act as an alternative H_2 sink and mitigate methanogenesis (Nisbet and Martin, 1993; Martin and Streeter, 1995; McAllister and Newbold, 2008). It has been reported by the studies that supplementation of fumaric acid or malic acid reduced CH_4 production (Callaway and Martin, 1996; McGinn et al., 2004; Foley et al., 2009; Wood et al., 2009). However, the authors stated that the effect of organic acid supplementation seems to be affected by diet. Greater reductions in methane production were obtained when diets were supplemented with high concentrate, most likely caused by a greater effect on the acetate to propionate (A:P) ratio in the rumen, in addition to its ability to function as a H_2 sink. High levels of organic acid supplementation decrease DMI and ruminal pH, adversely affecting fiber fermentation in the rumen (Beauchemin et al., 2009). This can be overcome by encapsulation of organic acids with fat in order to slow their release in the rumen (Wallace et al., 2006; Martin et al., 2010), but to date it has been considered expensive as a mitigation strategy.

7 Using cereal grains to improve feed efficiency and reduce methane production

In general, ruminants lose from 2% to 12% of their ingested energy as methane (Ferrell, 1988; Harper et al., 1999). Thus, strategies that can reduce emissions of methane from cattle are beneficial not only to the environment but also to the animal, as they normally result in improved animal performance if the C and H_2 are captured in a form available to the animal as metabolizable energy. If H_2 is simply captured in another form or lost as H_2 gas then it will still be lost to the animal. Simply reducing the populations of methanogens can reduce methane production (at least temporarily), but this methane reduction alone does not affect animal performance (Patra and Saxena, 2009; Hook et al., 2010; Wright and Klieve, 2011). Thus it is imperative that the energy lost as methane must be redirected into forms that are usable by the animal, such as propionate, so that more metabolizable energy reaches the animal. As has been noted above, production of methane by cattle is affected by numerous factors such as the addition of ionophores and lipids to diets, the level of feed intake, and even by the type of diet fed to the cattle (Johnson and Johnson, 1995; Grainger and Beauchemin, 2011). In this section, the authors focus on the latter factor: the type of diet offered to cattle. More specifically, the authors focus on the use of cereal grains and how their increased proportion of inclusion in the diet (at the expense of roughage sources) can affect feed efficiency and production of methane.

Fermentation of rapidly degradable carbohydrates (e.g. starch) in the rumen promotes production of propionate, creating an alternative hydrogen sink to methanogenesis (Grainger and Beauchemin, 2011). In addition, this type of fermentation lowers ruminal pH, inhibits the growth of rumen methanogens and decreases rumen protozoal numbers. It has been suggested that ruminal protozoa serve to sequester starch as they engulf and subsequently ferment starch granules. As further evidence of the role that protozoa play in ruminal starch fermentation, more Entodinium were associated with corn than barley diets (Xia et al., 2015). A reduction in the protozoa population also contributes to methane mitigation as it limits the transfer of hydrogen from protozoa to methanogens (Grainger and Beauchemin, 2011).

A study conducted by Harper and collaborators measured methane emissions from cattle under grazing and feedlot conditions. Among their findings, authors reported that when cattle were grazed on pasture, they wasted an equivalent of 7.7–8.4% of their GE intake as methane. However, when the same group of heifers was fed a highly digestible, high-grain diet, the amount of energy lost as methane decreased to only 1.9–2.2% of their GE intake. This difference of almost four times in magnitude demonstrates that cattle fed low-quality (high-fiber) diets produce more methane than cattle fed high-quality, high-grain diets (Harper et al., 1999). Similarly, in an effort to quantify the amount of methane produced at different ratios of forage to concentrate, Hales et al. (2014) fed diets varying from 2% to 14% alfalfa hay to beef steers. These authors detected a linear reduction in dry matter digestibility, digestible energy, and metabolizable energy as the level of forage increased (or as levels dry-rolled corn decreased) in the steers' diet. This resulted in lower energy retained as body weight gain. This lower retained energy was attributed to a couple of factors which included a higher loss of energy as methane, since steers had greater levels of forage in their diets. For instance, the percentage of energy intake lost as methane increased from 3.07% to 4.18% as the level of forage in the diet varied from 2% to 14%. The authors concluded that methane losses increased linearly as the level of forage increased in the diet, which coincided with less energy being retained by the animals.

Essentially, fermentation of cell wall components (cellulose and hemicellulose) results in higher acetate to propionate molar ratios, which leads to higher methane yields compared to the fermentation of soluble carbohydrates (Johnson and Johnson, 1995). Because the quantity of methane generated is related to the end products produced from carbohydrate fermentation in the rumen (Fahey and Berger, 1988), higher amounts of acetate at the expense of propionate results in higher methane production. As illustrated in Fig. 3, if a high-grain or a high-forage diet is fed to cattle, resulting in distinct acetate to propionate ratios, the same amount of glucose available for fermentation would result in very different quantities of methane being formed. In fact, for the two examples presented in Fig. 1 the amount of methane would be 1.8 times greater in the high-forage diet, compared to the high-grain diet. Thus, there is a positive relationship between acetate and methane, and a negative relationship between propionate and methane. According to Fahey and Berger (1988), these relationships stem from the fact that when the propionate is being formed, more C and H atoms present in glucose are accounted for. In contrast, more H atoms are released when glucose is converted to acetate, resulting in greater quantities of methane being formed.

About two-thirds of the methane produced in nature originate from acetate, more specifically, from the methyl group of acetate (Ferry, 1992; Wolfe, 1993). As previously stated, from the productivity stand point, methane is considered a waste of energy. Therefore, one of the reasons why animal performance is normally improved when more propionate is produced (instead of acetate) is because more C and H atoms become propionate instead of methane, increasing the metabolizable energy level of the diet (Fahey and Berger, 1988). In practical situations, the ratio of acetate to propionate normally varies from 0.9 to 4.0, and consequently, corresponding methane losses vary widely as well. However, as highlighted in this section, the literature clearly shows that feeding more grains to ruminants can substantially decrease their methane emissions. Not only can it lower methane emissions per kg of DM consumed, but it can also reduce total methane emissions per kg of final product (e.g. milk, meat) since more grains normally improve animal performance. Therefore, the development of strategies to mitigate emissions of methane by cattle is critical and achieving this goal while not reducing or even improving animal performance is highly desirable; therefore, increasing the levels of cereal grains fed to cattle 'kills two birds with one stone'.

Theoretical Input → Output Models

Hypothetical situation #1: High-grain diet
(theoretical acetate:propionate ratio = 1)

15 glucose → 10 acetate + 10 propionate + 5 butyrate + 15 CO_2 + 10 H_2O + **5 CH_4**

Versus

Hypothetical situation #2: High-forage diet
(theoretical acetate:propionate ratio = 3)

15 glucose → 18 acetate + 6 propionate + 3 butyrate + 15 CO_2 + 18 H_2O + **9 CH_4**

Figure 3 Theoretical models illustrating the ruminal fermentation of 15 molecules of glucose under two hypothetical situations: High-grain and high-forage diets.

8 Microbiology of cereal grain fermentation

The rumen microflora is a dense and diverse consortium of archaea, bacteria, fungi, and protozoa all competing for resources through the diet of their ruminant host. Uniquely, these microorganisms have given the ruminant animal an evolutionary niche in which the microbes are able to utilize complex carbohydrates, such as cellulose and hemicellulose (fibrolytic microbes), through certain metabolic pathways which yield fermented by-products that are absorbed and utilized by the animal. The ruminant host in return provides a warm (39°C), anaerobic, and nutrient-rich media for the microbes to thrive. Starch-fermenting microbes contain alpha-amylase and beta-amylase which allow them to degrade large polysaccharides of glucose in forms of amylose and amylopectin. Generally speaking, most starch and forage fermentation results in short-chain fatty acid production (formate, acetate, propionate, butyrate, and branch-chain SCFA), adenosine triphosphate production for microbial cell energy, microbial crude protein synthesis, and heat + gas (methane and carbon dioxide) as an energetic loss to the animal and microbes.

Drs. Robert Hungate and Marvin Bryant are the forebearers of modern ruminal microbiology and microbial ecology (Krause et al., 2013), and were two of the first ruminant microbiologists to study and isolate individual ruminal microbes in pure culture to further understand the symbiotic relationship which exists between microbes and the ruminant animal (Hungate, 1944, 1947; Bryant, 1959). Lastly, a ruminant animal relies on the majority of fermentation end products from foregut fermentation in the rumen, but just like in a monogastric digestive system, there is a second site of fermentation in the lower digestive tract of a ruminant which accounts for 10–20% of the energy for the animal (Russell, 2002). In vitro studies have shown that specific methane production from cecal digesta is 6.8% of the rumen digesta (Freetly et al., 2015). Furthermore, no differences in methanogenic bacterial populations were observed between the two sites of digestion. However, for the purposes of this publication the predominant starch utilizing microbes and methane-producing archaea found in the rumen will be discussed in full.

The early work of Hungate and Bryant led to the general classifications of microbes into primary niche categories: fibrolytic, simple carbohydrate, obligate amino acid, and methane-producing fermenting microorganisms. Although it is easy to divide these ruminal microbes into categories, it is important to note that current technology has indicated that the microbial diversity may be larger than scientists have previously recognized. The microbial ecology of the rumen is very rich, and a variety of starch-fermenting bacteria and methane-producing archaea are known because they have all been grown and studied in pure culture experimentations. Table 1 below illustrates the niches and by-products of the starch-degrading and methane-producing microbes modified from Russell (2002).

As research has moved from culture-based methodologies to Next-Generation Sequencing and the Hungate 1000 project data, further information about changes in the microbial population have emerged (McAllister et al., 2015; Seshadri et al., 2018). Using starch as a selective agent, ruminal fluid cultures had increased *Prevotella* populations associated with starch feeding (Bandarupalli, 2017). However many of the bacterial species thought to be responsible for most of the starch fermentation in the rumen based on in vitro studies (e.g. *Streptococus bovis*, *Ruminobacter amylophilus*, *Succinomonas amylolytica*, *Butyrivibrio fibrisolvens*) were not detected at high populations in cattle-fed corn or barley diets (Xia et al., 2015). Cattle with frothy bloat from grain rations had increased populations of *Clostridium*, *Eubacterium*, and *Butyrivibrio*, while having reduced *Prevotella* populations (Pitta et al., 2016).

Table 1 Ruminal species and their niche(s) and by-products

Species	Primary niches	Products
Butyrivibrio fibrisolvens	cellulose, hemicellulose, starch, pectin, and sugar	butyrate, formate, lactate, and acetate
Ruminobacter amylophilus	starch	succinate, formate, and acetate
Selenomonas ruminantium	sugar, starch, and lactate	lactate, acetate, propionate, butyrate, and H_2
Prevotella ruminicola, P. albensis, P. bryantii, P. brevis	starch, hemicellulose, pectin, -glucans, protein	succinate, acetate, formate, propionate
Succinomonas amylolytica	starch	succinate, acetate, propionate
Streptococcus bovis	starch, sugar	lactate, acetate, formate, ethanol
Methanobrevibacter ruminantium	H_2, carbon dioxide, formate	methane
Methanosphaera stadtmaniae[a]	H_2, methanol	methane

[a] Information is cited from Liu and Whitman (2008).
Source: modified from Russell (2002).

9 Bacteria and archaea involved in fermentation

9.1 Butyrivibrio fibrisolvens

Butyrivibrio sp. are versatile, gram-negative, rod-shaped microbes which have enzyme capabilities of degrading pentose and hexose sugars, along with starch and hemicellulose. Bryant first isolated the microbe and classified this genus appropriately based on its characteristic butyrate production: hexose, pyruvate, acetyl CoA, butyrate (NADP+ regeneration), and acetate (ATP generation). Some *Butyrivibrio* sp. have butyrate kinase and will never generate lactate; however, the *Butyrivibrio* sp. that do not have butyrate kinase will revert carbon skeletons to lactate when acetate is absent.

9.2 Ruminobacter amylophilus

The ruminal bacteria *R. amylophilus* is a great model for starch degradation because it only uses 1–4 polymers of glucose for energy, and it was the first reported ruminal bacterium demonstrated to exhibit starch-binding sites on the cell surface similar to the intestinal *Bacteroides* (Anderson, 1995). *R. amylophilus* is a gram-negative, anaerobic, and proteolytic (only with ammonia nitrogen) microbe that relies on polysaccharide degradation of maltose, maltodextrins, and starch.

9.3 Selenomonas ruminantium

Selenomonas ruminantium is a gram-negative and strictly anaerobic bacterium with a rod-shaped structure and unique flagella that allows the microbe to turn/spin on both axes. *S. ruminantium* can grow very rapidly in the presence of sugar, and ferments almost strictly in a homolactic manner (pyruvate to lactic acid). Although *Sel. ruminantium*'s niche is related to monosaccharides or disaccharides, in the absence of simple sugars *Sel. ruminantium* can utilize lactic acid and dextrins from other starch-fermenting microbes as its main source of ATP generation.

9.4 Prevotella sp.

Prevotella ruminicola was the first microbe to be reclassified from the *Bacteroides ruminicola* classification because of its sensitivity to bile salts and hexose monophosphate pathway (Russell, 2002). Around the mid-1990s, *P. ruminicola* was then further reclassified into a variety of species (*ruminicola, albensis, bryantii,* and *brevis*) based on the enzymatic capabilities and xylose utilization: carboxymethyl cellulase negative, produced deoxyribonuclease, and no xylose utilization; *Prevotella brevis,* carboxymethyl cellulase positive, produced deoxyribonuclease, and utilized xylose; *Prevotella bryantii,* and carboxymethyl cellulase positive and did not produce deoxyribonuclease; *Prevotella ruminicola* (Avgustin et al., 1997). *Prevotella sp.* are pleomorphic rod-shaped bacteria which generate a large amount of succinate, utilize multiple substrates for energy except for cellulose, and are in typically the most abundant genus within the rumen.

9.5 Succinomonas amylolytica

Succinomonas amylolytica is a strict anaerobe, rod to coccoid shaped cell that occupies a starch-degrading niche in the rumen (Bryant et al., 1957). *Suc. amylolytica* grows well in cultures rich in trypticase and yeast when rumen fluid is removed but does not grow at all with no bicarbonate as indicated. *Suc. amylolytica* produces little amounts of propionate compared to its major by-products: succinate and acetate.

9.6 Streptococcus bovis

Streptococcus bovis is a gram-positive, facultative anaerobe which grows the fastest in reduced environments in the absence of oxygen, and is ovoid in shape. *Str. bovis* is one of the fastest growing bacteria in the rumen with a published doubling time as short as 24 min (Russell and Robinson, 1984). However, there are unfortunate repercussions of this fast doubling time because *Str. bovis'* homolactic fermentations produce excessive amounts of lactic acid when sugars and starches are readily available which results in decrease in the pH and an increase in the acidotic conditions for ruminant animals.

9.7 Archaea involved in fermentation

Methanobrevibacter ruminantium and *Methanosphaera stadtmanae,* both belong to the archaea family Methanobacteriaceae. Methane is a major by-product of ruminal fermentation and can produce as much as roughly 400 liters from mature cattle in a day. Archaea are widely distributed in cattle, and *Methanobrevibacter gottschalkii* and *Methanobrevibacter ruminantium* accounted for up to 74% of all archaea (Henderson et al., 2015). Five dominant methanogen groups comprised more than 89% of the archeal communities (Henderson et al., 2015).

Ruminant microbiologists and nutritionists have investigated alternatives to reduce methane production because as much as 10% of dietary GE is lost through methane production (Blaxter, 1962). However, methanogenesis is a natural process in the rumen and is needed to maintain low hydrogen-ion concentrations by methanogens which utilize H^+ and CO_2 and produce methane as a by-product. Methane-producing bacteria tend to have a stronger symbiotic relationship and effectiveness with fibrolytic microbes versus starch-degrading bacteria. It has long been understood that when cereal grains are added to a rumen diet, the propionate increases and methane production decreases (Van Kessel and Russell, 1996; Czerkawski, 1986). Frothy bloat is a disease encountered in many feedlot cattle as a result of gas being captured in the viscous ruminal fluid of cattle fed higher grain diets, and archeael populations (most notably *Methanobrevibacter*) were higher in bloated cattle compared to controls (Pitta et al., 2016) (Fig. 4).

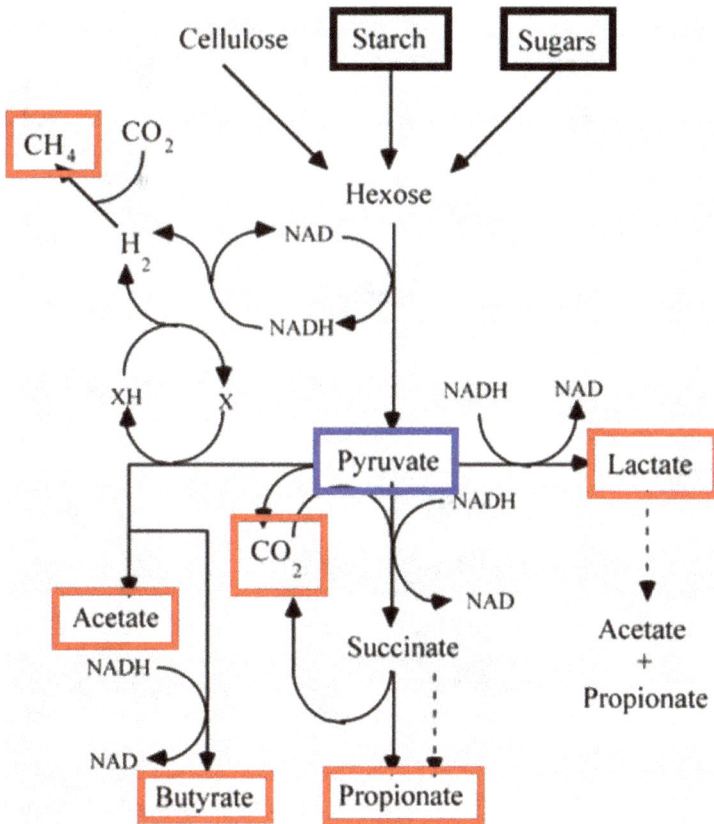

Figure 4 The black boxes indicate simple carbohydrate substrates in ruminant diets. The red boxes indicate end products of microbial fermentation. The blue box highlights pyruvate, which is the pivot point/intermediate for several microbial pathways. Source: modified from Russell and Rychlik (2001).

10 Feed retention time

One of the factors that contributes most to degradation of feedstuffs and absorption of nutrients is the duration feedstuffs are retained in the rumen before passing through the reticulo-omasal orifice and into the rest of the GIT. The retention time of feedstuffs in the rumen varies depending on a variety of factors, one of which being the diet fed to the animal. Concentrate diets high in readily fermentable starches decrease the retention time, as they are broken down rapidly. Increasing the dilution rate in a RUSITEC (rumen simulation technique) fermenter with ovine rumen fluid inoculum showed an increase in DMI as well as VFA and ammonia-N production (Martínez et al., 2009).

When retention time of feedstuffs in the rumen decreases, the rate of methane production by the methanogens in the rumen also decreases because methanogens have a relatively slow growth rate and are unable to maintain adequate numbers to sustain greater methane production. An increase in the dilution rate of rumen fluid can inhibit the growth of methanogens, limiting their abundance in the rumen (Eun et al., 2004) and ultimately leading to a reduction in the methane they are able to produce. This effect can be seen by altering the diet or other factors. Okine et al. (1989) found that displacing free volume in the rumen with inert materials, methane production decreased 29% along with an increase in rumen fluid dilution rate of 43%. A quarter of the variation in methane production by a rumen could be explained by how quickly the fluid is washed out of the rumen during and after feeding (Okine et al., 1989).

An animal can also be genetically selected for a particular level of methane production, and they have similar patterns between methane production and rumen retention time. Sheep producing less methane have a decrease in the retention times of both the solid and liquid dilution rates accounting for 59% and 70% of the variation in methane production, respectively (Goopy et al., 2014). Sheep that produced less methane also had a lower amount of particulate (g DM) in their rumen than sheep producing more methane (Goopy et al., 2014). A positive relationship exists between methane production and rumen liquid and particulate mean retention times. As noted numerous times above, methane production is correlated to ruminal retention time.

11 Acidosis and other negative feed effects

If too much grain is introduced into the diet too quickly, the pH of the rumen can drop rapidly, causing ruminal acidosis. Acidosis can be either acute or subacute (chronic). Acute acidosis occurs when the pH of the rumen drops abruptly after cattle consume a lot of readily fermentable starch. Subacute acidosis occurs when the pH of the rumen is consistently too low (pH between 5.1 and 5.6) which can happen in feedlot settings where cattle are pushed to their limits when it comes to concentrates in the diet (Brown et al., 2006). Subacute acidosis does not manifest itself into observable symptoms and the cattle may not even appear to be sick until death occurs (Owens et al., 1998). Not all animals are affected the same by adding grain to the diets. On the same diet, some cattle have shown severe adverse effects, and some have only shown mild effects such as diarrhea (Brown et al., 2006). Regardless of severity, treatment is needed because it can cause animals to either reduce their DMI or go 'off-feed' completely. In order to not suddenly and drastically disrupt the microbial populations within the ruminal environment, using a step-up ration can gradually allow the ruminal microbial population to adapt and reduce abundance of *Str. bovis* which can overgrow, rapidly produce lactic acid, and decrease the pH of the rumen (Wells et al., 1997). This helps the microbes in the rumen adapt to the new environment, so they are better suited for it.

A drop in ruminal pH cause risk for acidosis but itcan also damage the lining of the rumen epithelium causing keratinization or parakeratosis which is a hardening of the ruminal epithelial cells used for absorption. A diet with 60% grain showed an increase in the ruminal epithelial cell thickening in goats when compared to 0% and 30% grain diets. This damage ultimately leads to a decrease in absorption of nutrients and VFAs from the rumen because they cannot travel across the epithelial cells into the bloodstream to be absorbed by the animal (Hinders and Owen, 1965). The lack of absorption causes a decrease in the energy availability to the animal and negatively impacts the overall production of the animal.

Not only does feeding high-grain diets decrease the pH and damage the lining of the rumen, there is also an increase in the prevalence liver abscesses. When the ruminal epithelium becomes damaged by low pH associated with feeding starch, pathogens have an opportunity to colonize the rumen. An example is the anaerobic bacteria *Fusobacterium necrophorum*, which is the main cause of liver abscesses (Nagaraja and Chengappa, 1998). After colonization, the bacteria can enter the blood stream through the damaged epithelium and eventually end up being filtered by the liver where it can cause abscesses (Nagaraja and Chengappa, 1998). This process can occur during all levels of production in both dairy and beef operations; however, it is most common in feedlots when the cattle are placed on a high-concentrate diet.

Liver abscesses are a major concern in the feedlot system with 12–32% of young feedlot cattle exhibiting significant liver damage at slaughter (Brink et al., 1990). This causes economic loss for packers because they cannot sell the liver and must also spend more time trimming the carcass. Not only do liver abscesses cause the liver to be condemned, but the animal's prior performance also gets affected. Cattle with liver abscesses have a lower feed intake and less gain (Brink et al., 1990). A preventive measure in the past, the antibiotic tylosin phosphate has been included in the diet to reduce ruminal *F. necrophorum*, prevent liver abscesses, and increase weight gain (Brown et al., 1975).

12 Summary

The use of starch-containing cereal grains as a feedstuff improves the efficiency because these energy-dense feedstuffs are more fermentable than forages and the fermentation of starch generally results in a shift in VFAs and less GHGs. The reduction of the reducing equivalent sink (CH_4) for disposal of accumulating NADH from catabolic reactions further ensures an increase in the production of propionate, a reduced VFA. Because propionate is a glucogenic VFA, it yields more ATP when metabolized by the ruminant animal and is also associated with higher levels of intramuscular marbling. Furthermore, because the grain shifts the fermentation profile away from acetate production, there is a concomitant reduction in CO_2 production by the ruminal carbohydrate fermentation, which ensures an improvement in sustainability of animal production. Thus, feeding ruminants cereal grains has profound impact on animal physiology as well as efficiency, and reduces the environmental footprint of ruminant animal production.

Feeding high-starch diets to cattle has many benefits to carcass quality and growth efficiency in addition to reducing methane production. However, there are still negative consequences to grain feeding, including lowering of pH and inducing acidosis in cattle, as well as liver abscesses. High-starch rations have benefits to producers and environmental sustainability that are important to maintain, but the biochemical basis of these benefits must be more well-characterized. The impact of starch feeding on the composition of the microbial population of the rumen and lower GIT is profound, and is related to the efficiency as well as the sustainability of beef production. Until these benefits derived from starch feeding can be replicated via other methods, then feeding high-grain rations will be favored, when viewed from both the economic and environmental lenses.

13 Where to look for further information

The topic of grain feeding in cattle is inextricably enmeshed with current discussions of environmental impact and long-term production sustainability. However, this present chapter is focused primarily on the impact within the animal, and what effects these diets have on the microbial population. We must understand the succession of events that occur in the ruminal microbial population when cattle are fed high-starch rations. If we can fully understand the shifts in the microbiome and metabolome of the rumen, then we can begin to prepare specific interventions or remedies to ameliorate the negative impacts (such as acidosis, both acute and sub-acute) and reduce production of wasteful end products (e.g. methane and CO_2) of fermentation. Furthermore, we must understand the effect of other least-cost feed components on the microbial populations as well and their impacts on animal health, immunity, and food safety.

There are many organizations involved in research that are aimed at furthering our understanding of the impact of high-grain rations on the rumen ecosystem and physiology. In the United States, the Agricultural Research Service, most notably at the Meat Animal Research Center, along with the Rowett Research Institute in Aberdeen (now a part of University of Aberdeen) and the Roslin Institute, as well as INRA in France, and of course CSIRO in Australia. Many university researchers around the world have been involved in expanding our understanding, frankly too many to list here because of the quality scientists that would inevitably be inadvertently left out of the list. Research in this arena is highly active and ongoing, and new discoveries are made yearly Many of the most active researchers in this field attend the Congress on Gastrointestinal Function (https://www.co ngressgastrofunction.org/) as well as the companion INRA-Aberdeen Joint Symposium on Gastrointestinal Microbiology. One of the best new resources available to examine many of the issues described here is: Tedeschi and Nagaraja (2020).

14 References

Anderson, K. L. 1995. Biochemical analysis of starch degradation by *Ruminobacter amylophilus* 70. *Applied and Environmental Microbiology* 61(4), 1488–91. doi:10.1128/AEM.61.4.1488-1491.1995.
Archimède, H., Eugène, M., Marie Magdeleine, C., Boval, M., Martin, C., Morgavi, D. P., Lecomte, P. and Doreau, M. 2011. Comparison of methane production between C3 and C4 grasses and legumes. *Animal Feed Science and Technology* 166–167, 59–64. doi:10.1016/j.anifeedsci.2011.04.003.

Avgustin, G., Wallace, R. J. and Flint, H. J. 1997. Phenotypic diversity among ruminal isolates of *Prevotella ruminicola*: proposal of *Prevotella brevis* sp. nov., *Prevotella bryantii* sp. nov., and *Prevotella albensis* sp. nov. and redefinition of *Prevotella ruminicola*. *International Journal of Systematic Bacteriology* 47(2), 284–8. doi:10.1099/00207713-47-2-284.

Awika, J. M. 2011. Major cereal grains production and use around the world. *ACS Symposium Series* 1089(1), 1–13. doi:10.1021/bk-2011-1089.ch001.

Bandarupalli, V. 2017. Identification of novel rumen bacteria using starch as a selective nutrient in batch cultures. *Journal of Animal Science* 95(suppl_4), 305. doi:10.2527/asasann.2017.623.

Beauchemin, K. A., McAllister, T. A., Dong, Y., Farr, B. I. and Cheng, K. J. 1994. Effects of mastication on digestion of whole cereal grains by cattle. *Journal of Animal Science* 72(1), 236–46. doi:10.2527/1994.721236x.

Beauchemin, K. A., Kreuzer, M., O'mara, F. and McAllister, T. A. 2008. Nutritional management for enteric methane abatement: a review. *Australian Journal of Experimental Agriculture* 48(2), 21–7. doi:10.1071/EA07199.

Beauchemin, K. A., McAllister, T. A. and McGinn, S. M. 2009. Dietary mitigation of enteric methane from cattle. *CAB Reviews: Perspectives in Agriculture, Veterinary Science, Nutrition and Natural Resources* 4(35), 1–18. doi:10.1079/PAVSNNR20094035.

Beever, D. E., Dhanoa, M. S., Losada, H. R., Evans, R. T., Cammell, S. B. and France, J. 1986. The effect of forage species and stage of harvest on the processes of digestion occurring in the rumen of cattle. *The British Journal of Nutrition* 56(2), 439–54. doi:10.1079/bjn19860124.

Benchaar, C., Pomar, C. and Chiquette, J. 2001. Evaluation of dietary strategies to reduce methane production in ruminants: a modelling approach. *Canadian Journal of Animal Science* 81(4), 563–74. doi:10.4141/A00-119.

Bergen, W. G. and Bates, D. B. 1984. Ionophores: their effect on production efficiency and mode of action. *Journal of Animal Science* 58(6), 1465–83. doi:10.2527/jas1984.5861465x.

Blaxter, K. 1962. The energy metabolism of ruminants. In: Blaxter, K. (Ed.), *The Energy Metabolism of Ruminants*. Charles C. Thomas, Springfield, IL, pp. 197–200.

Boadi, D., Benchaar, C., Chiquette, J. and Massao, D. 2004. Mitigation strategies to reduce enteric methane emissions from dairy cows: update review. *Canadian Journal of Animal Science* 84(3), 319–35. doi:10.4141/A03-109.

Brink, D. R., Lowry, S. R., Stock, R. A. and Parrott, J. C. 1990. Severity of liver-abscesses and efficiency of feed-utilization of feedlot cattle. *Journal of Animal Science* 68(5), 1201–7. doi:10.2527/1990.6851201x.

Brown, H., Bing, R. F., Grueter, H. P., McAskill, J. W., Cooley, C. O. and Rathmacher, R. P. 1975. Tylosin and chlortetracycline for the prevention of liver abscesses, improved weight gains and feed efficiency in feedlot cattle. *Journal of Animal Science* 40(2), 207–13. doi:10.2527/jas1975.402207x.

Brown, M. S., Ponce, C. H. and Pulikanti, R. 2006. Adaptation of beef cattle to high-concentrate diets: performance and ruminal metabolism. *Journal of Animal Science* 84(Suppl.), E25–33. doi:10.2527/2006.8413_supple25x.

Bryant, M. P. 1959. Bacterial species of the rumen. *Bacteriological Reviews* 23(3), 125–53. doi:10.1128/MMBR.23.3.125-153.1959.

Bryant, M. P., Small, N., Bouma, C. and Chu, H. 1957. *Bacteriodes ruminicola* N. Sp. and *Succinimonas amylolytica* the new genus and species. *Journal of Bacteriology* 76, 15–23.

Buddle, B. M., Denis, M., Attwood, G. T., Altermann, E., Janssen, P. H., Ronimus, R. S., Pinares-Patiño, C. S., Muetzel, S. and Wedlock, D. N. 2011. Strategies to reduce methane emissions from farmed ruminants grazing on pasture. *The Veterinary Journal* 188(1), 11–7. doi:10.1016/j.tvjl.2010.02.019.

Callaway, T. R. and Martin, S. A. 1996. Effects of organic acid and monensin treatment on in vitro mixed ruminal microorganism fermentation of cracked corn. *Journal of Animal Science* 74(8), 1982–9. doi:10.2527/1996.7481982x.

Czerkawski, J. W. 1986. *An Introduction to Rumen Studies*. Elsevier Press, Oxford, UK.

Depenbusch, B. E., Nagaraja, T. G., Sargeant, J. M., Drouillard, J. S., Loe, E. R. and Corrigan, M. E. 2008. Influence of processed grains on fecal pH, starch concentration, and shedding of *Escherichia coli* O157 in feedlot cattle. *Journal of Animal Science* 86(3), 632–9. doi:10.2527/jas.2007-0057.

Dong, H., Mangino, J., McAllister, T. A., Hatfield, J. L., Johnson, D. E., Lassey, K. R., Lima, M. A. d. and Romanovskaya, A. 2006. Emissions from livestock and manure management. In: Eggleston, S., Buendia, L., Miwa, K., Ngara, T. and Tanabe, K. (Eds), *Guidelines for National Greenhouse Gas Inventories* (vol. 4). Institute for Global Environmental Strategies, Hayama, Japan, pp. 10.11–87.

Eckard, R. J., Grainger, C. and De Klein, C. A. M. 2010. Options for the abatement of methane and nitrous oxide from ruminant production: a review. *Livestock Science* 130(1–3), 47–56. doi:10.1016/j.livsci.2010.02.010.

Ellis, J. L., Dijkstra, J., Kebreab, E., Bannink, A., Odongo, N. E., McBride, B. W. and France, J. 2008. Aspects of rumen microbiology central to mechanistic modelling of methane production in cattle. *The Journal of Agricultural Science* 146(2), 213–33. doi:10.1017/S0021859608007752.

Etuk, E., Ifeduba, A., Okata, U., Chiaka, I., Okoli, I. C., Okeudo, N., Esonu, B., Udedibie, A. and Moreki, J. 2012. Nutrient composition and feeding value of sorghum for livestock and poultry: a review. *Journal of Animal Science Advances* 2(6), 510–24.

Eun, J. S., Fellner, V. and Gumpertz, M. L. 2004. Methane production by mixed ruminal cultures incubated in dual-flow fermentors. *Journal of Dairy Science* 87(1), 112–21. doi:10.3168/jds.S0022-0302(04)73148-3.

Fahey, G. C. and Berger, L. L. 1988. Carbohydrate nutrition of ruminants. In: Church, D. C. (Ed.), *The Ruminant Animal: Digestive Physiology and Nutrition*. Prentice Hall, Englewood Cliffs, NJ, pp. 269–97.

Ferrell, C. L. 1988. Energy metabolism. In: Church, D. C. (Ed.), *The Ruminant Animal: Digestive Physiology and Nutrition*. Prentice Hall, Englewood Cliffs, NJ.

Ferry, J. G. 1992. Methane from acetate. *Journal of Bacteriology* 174(17), 5489–95. doi:10.1128/jb.174.17.5489-5495.1992.

Firkins, J. L., Berger, L. L. and Fahey Jr., G. C. 1985. Evaluation of wet and dry distillers grains and wet and dry corn gluten feeds for ruminants. *Journal of Animal Science* 60(3), 847–60. doi:10.2527/jas1985.603847x.

Foley, P. A., Kenny, D. A., Callan, J. J., Boland, T. M. and O'Mara, F. P. 2009. Effect of DL-malic acid supplementation on feed intake, methane emission, and rumen fermentation in beef cattle. *Journal of Animal Science* 87(3), 1048–57. doi:10.2527/jas.2008-1026.

Forster, P., Ramaswamy, V., Artaxo, P., Berntsen, T., Betts, R., Fahey, D. W., Haywood, J., Lean, J., Lowe, D. C., Myhre, G., Nganga, J., Prinn, R., Raga, G., Schulz, M. and Van Dorland, R. 2007. Changes in atmospheric constituents and in radiative forcing. In: Solomon, S., Qin, D., Manning, M., Chen, Z., Marquis, M., Averyt, K. B., Tignor, M. and Miller, H. L. (Eds), *Climate Change 2007: The Physical Science Basis. Contribution of Working Group I to the Fourth Assessment Report of the Intergovernmental Panel on Climate Change*. Cambridge University Press, Cambridge, UK and New York, NY.

Freetly, H. C., Lindholm-Perry, A. K., Hales, K. E., Brown-Brandl, T. M., Kim, M., Myer, P. R. and Wells, J. E. 2015. Methane production and methanogen levels in steers that differ in residual gain. *Journal of Animal Science* 93(5), 2375–81. doi:10.2527/jas.2014-8721.

Fron, M., Madeira, H., Richards, C. and Morrison, M. 1996. The impact of feeding condensed distillers byproducts on rumen microbiology and metabolism. *Animal Feed Science and Technology* 61(1–4), 235–45. doi:10.1016/0377-8401(95)00943-4.

Gerber, P. J., Steinfeld, H., Henderson, B., Mottet, A., Opio, C., Dijkman, J., Falcucci, A. and Tempio, G. 2013. *Tackling Climate Change through Livestock: a Global Assessment of Emissions and Mitigation Opportunities*. Food and Agriculture Organization of the United Nations (FAO), Rome.

Gómez, L. M., Posada, S. L. and Olivera, M. 2016. Starch in ruminant diets: a review. *Revista Colombiana de Ciencias Pecuarias* 29(2), 77–90. doi:10.17533/udea.rccp.v29n2a01.

Goopy, J. P., Donaldson, A., Hegarty, R., Vercoe, P. E., Haynes, F., Barnett, M. and Oddy, V. H. 2014. Low-methane yield sheep have smaller rumens and shorter rumen retention time. *The British Journal of Nutrition* 111(4), 578–85. doi:10.1017/S0007114513002936.

Grainger, C. and Beauchemin, K. A. 2011. Can enteric methane emissions from ruminants be lowered without lowering their production? *Animal Feed Science and Technology* 166–167, 308–20. doi:10.1016/j.anifeedsci.2011.04.021.

Guan, H., Wittenberg, K. M., Ominski, K. H. and Krause, D. O. 2006. Efficacy of ionophores in cattle diets for mitigation of enteric methane. *Journal of Animal Science* 84(7), 1896–906. doi:10.2527/jas.2005-652.

Hales, K. E., Brown-Brandl, T. M. and Freetly, H. C. 2014. Effects of decreased dietary roughage concentration on energy metabolism and nutrient balance in finishing beef cattle. *Journal of Animal Science* 92(1), 264–71. doi:10.2527/jas.2013-6994.

Hammond, K. J., Burke, J. L., Koolaard, J. P., Muetzel, S., Pinares-Patiño, C. S. and Waghorn, G. C. 2013. Effects of feed intake on enteric methane emissions from sheep fed fresh white clover (*Trifolium repens*) and perennial ryegrass (*Lolium perenne*) forages. *Animal Feed Science and Technology* 179(1–4), 121–32. doi:10.1016/j.anifeedsci.2012.11.004.

Haque, M. N. 2018. Dietary manipulation: a sustainable way to mitigate methane emissions from ruminants. *Journal of Animal Science and Technology* 60(1), 15. doi:10.1186/s40781-018-0175-7.

Harper, L. A., Denmead, O. T., Freney, J. R. and Byers, F. M. 1999. Direct measurements of methane emissions from grazing and feedlot cattle. *Journal of Animal Science* 77(6), 1392–401. doi:10.2527/1999.7761392x.

Henderson, G., Cox, F., Ganesh, S., Jonker, A., Young, W., Janssen, P. H., Abecia, L., Angarita, E., Aravena, P., Arenas, G. N., Ariza, C., Attwood, G. T., Avila, J. M., Avila-Stagno, J., Bannink, A., Barahona, R., Batistotti, M., Bertelsen, M. F., Brown-Kav, A., Carvajal, A. M., Cersosimo, L., Chaves, A. V., Church, J., Clipson, N., Cobos-Peralta, M. A., Cookson, A. L., Cravero, S., Carballo, O. C., Crosley, K., Cruz, G., Cucchi, M. C., De La Barra, R., De Menezes, A. B., Detmann, E., Dieho, K., Dijkstra, J., Dos Reis, W. L. S., Dugan, M. E. R., Ebrahimi, S. H., Eythórsdóttir, E., Fon, F. N., Fraga, M., Franco, F., Friedeman, C., Fukuma, N., Gagi , D., Gangnat, I., Grilli, D. J., Guan, L. L., Miri, V. H., Hernandez-Sanabria, E., Gomez, A. X. I., Isah, O. A., Ishaq, S., Jami, E., Jelincic, J., Kantanen, J., Kelly, W. J., Kim, S. H., Klieve, A., Kobayashi, Y., Koike, S., Kopecny, J., Kristensen, T. N., Krizsan, S. J., LaChance, H., Lachman, M., Lamberson, W. R., Lambie, S., Lassen, J., Leahy, S. C., Lee, S. S., Leiber, F., Lewis, E., Lin, B., Lira, R., Lund, P., Macipe, E., Mamuad, L. L., Mantovani, H. C., Marcoppido, G. A., Márquez, C., Martin, C., Martinez, G., Martinez, M. E., Mayorga, O. L., McAllister, T. A., McSweeney, C., Mestre, L., Minnee, E., Mitsumori, M., Mizrahi, I., Molina, I., Muenger, A., Munoz, C., Murovec, B., Newbold, J., Nsereko, V., O'Donovan, M., Okunade, S., O'Neill, B., Ospina, S., Ouwerkerk, D., Parra, D., Pereira, L. G. R., Pinares-Patino, C., Pope, P. B., Poulsen, M., Rodehutscord, M., Rodriguez, T., Saito, K., Sales, F., Sauer, C., Shingfield, K., Shoji, N., Simunek, J., Stojanovi - Radi , Z., Stres, B., Sun, X., Swartz, J., Tan, Z. L., Tapio, I., Taxis, T. M., Tomkins, N., Ungerfeld, E., Valizadeh, R., Van Adrichem, P., Van Hamme, J., Van Hoven, W., Waghorn, G., Wallace, R. J., Wang, M., Waters, S. M., Keogh, K., Witzig, M., Wright, A. D. G., Yamano, H., Yan, T., Yanez-Ruiz, D. R., Yeoman, C. J., Zambrano, R., Zeitz, J., Zhou, M., Zhou, H. W., Zou, C. X. and Zunino, P. 2015. Rumen microbial community composition varies with diet, host, but a core microbiome is found across a wide geographical range. *Scientific Reports* 5.

Herrera-Saldana, R. E., Huber, J. T. and Poore, M. H. 1990. Dry matter, crude protein, and starch degradability of five cereal grains. *Journal of Dairy Science* 73(9), 2386–93. doi:10.3168/jds.S0022-0302(90)78922-9.

Hinders, R. G. and Owen, F. G. 1965. Relation of ruminal parakeratosis development to volatile fatty acid absorption. *Journal of Dairy Science* 48(8), 1069–73. doi:10.3168/jds.S0022-0302(65)88393-X.

Hindrichsen, I. K., Wettstein, H. -R., Machmüller, A. and Kreuzer, M. 2006. Methane emission, nutrient degradation and nitrogen turnover in dairy cows and their slurry at different milk production scenarios with and without concentrate supplementation. *Agriculture, Ecosystems and Environment* 113(1–4), 150–61. doi:10.1016/j.agee.2005.09.004.

Holter, J. B. and Young, A. J. 1992. Methane prediction in dry and lactating Holstein cows. *Journal of Dairy Science* 75(8), 2165–75. doi:10.3168/jds.S0022-0302(92)77976-4.

Hook, S. E., Wright, A. D. and McBride, B. W. 2010. Methanogens: methane producers of the rumen and mitigation strategies. *Archaea* 2010, 945785. doi:10.1155/2010/945785.

Huhtanen, P. and Sveinbjörnsson, J. 2006. Evaluation of methods for estimating starch digestibility and digestion kinetics in ruminants. *Animal Feed Science and Technology* 130(1–2), 95–113. doi:10.1016/j.anifeedsci.2006.01.021.

Hungate, R. E. 1944. Studies on cellulose fermentation. I: The culture and physiology of an anaerobic cellulose-digesting bacterium. *Journal of Bacteriology* 48(5), 499–513. doi:10.1128/JB.48.5.499-513.1944.

Hungate, R. E. 1947. Studies on cellulose fermentation. III: The culture and isolation of cellulose-decomposing bacteria from the rumen of cattle. *Journal of Bacteriology* 53(5), 631–45. doi:10.1128/JB.53.5.631-645.1947.

Hungate, R. E. 1966. *The Rumen and Its Microbes.* Academic Press, New York, NY.

Huntington, G. B. 1997. Starch utilization by ruminants: from basics to the bunk. *Journal of Animal Science* 75(3), 852–67. doi:10.2527/1997.753852x.

Iannotti, E. L., Kafkewitz, D., Wolin, M. J. and Bryant, M. P. 1973. Glucose fermentation products of *Ruminococcus albus* grown in continuous culture with *Vibrio succinogenes*: changes caused by interspecies transfer of H_2. *Journal of Bacteriology* 114(3), 1231–40. doi:10.1128/JB.114.3.1231-1240.1973.

Johnson, K. A. and Johnson, D. E. 1995. Methane emissions from cattle. *Journal of Animal Science* 73(8), 2483–92. doi:10.2527/1995.7382483x.

Kebreab, E., Clark, K., Wagner-Riddle, C. and France, J. 2006. Methane and nitrous oxide emissions from Canadian animal agriculture: a review. *Canadian Journal of Animal Science* 86(2), 135–57. doi:10.4141/A05-010.

Kobayashi, Y. 2010. Abatement of methane production from ruminants: trends in the manipulation of rumen fermentation. *Asian-Australasian Journal of Animal Sciences* 23(3), 410–6. doi:10.5713/ajas.2010.r.01.

Koenig, K. M., Beauchemin, K. A. and Rode, L. M. 2003. Effect of grain processing and silage on microbial protein synthesis and nutrient digestibility in beef cattle fed barley-based diets. *Journal of Animal Science* 81(4), 1057–67. doi:10.2527/2003.8141057x.

Krause, D. O., Nagaraja, T. G., Wright, A. D. G. and Callaway, T. R. 2013. Board-invited review: rumen microbiology: leading the way in microbial ecology. *Journal of Animal Science* 91(1), 331–41. doi:10.2527/jas.2012-5567.

Kreikemeier, K. K., Harmon, D. L., Brandt Jr., R. T., Nagaraja, T. G. and Cochran, R. C. 1990. Steam-rolled wheat diets for finishing cattle: effects of dietary roughage and feed intake on finishing steer performance and ruminal metabolism. *Journal of Animal Science* 68(7), 2130–41. doi:10.2527/1990.6872130x.

Larsen, M., Lund, P., Weisbjerg, M. R. and Hvelplund, T. 2009. Digestion site of starch from cereals and legumes in lactating dairy cows. *Animal Feed Science and Technology* 153(3–4), 236–48. doi:10.1016/j.anifeedsci.2009.06.017.

Liu, Y. and Whitman, W. B. 2008. Metabolic, phylogenetic, and ecological diversity of the methanogenic Archaea. *Annals of the New York Academy of Sciences* 1125(1), 171–89. doi:10.1196/annals.1419.019.

Martin, S. A. and Streeter, M. N. 1995. Effect of malate on *in vitro* mixed ruminal microorganism fermentation. *Journal of Animal Science* 73(7), 2141–5. doi:10.2527/1995.7372141x.

Martin, C., Rouel, J., Jouany, J. P., Doreau, M. and Chilliard, Y. 2008. Methane output and diet digestibility in response to feeding dairy cows crude linseed, extruded linseed, or linseed oil. *Journal of Animal Science* 86(10), 2642–50. doi:10.2527/jas.2007-0774.

Martin, C., Morgavi, D. P. and Doreau, M. 2010. Methane mitigation in ruminants: from microbe to the farm scale. *Animal: an International Journal of Animal Bioscience* 4(3), 351–65. doi:10.1017/S1751731109990620.

Martínez, M. E., Ranilla, M. J., Ramos, S., Tejido, M. L. and Carro, M. D. 2009. Effects of dilution rate and retention time of concentrate on efficiency of microbial growth, methane production, and ruminal fermentation in *Rusitec fermenters*. *Journal of Dairy Science* 92(8), 3930–8. doi:10.3168/jds.2008-1975.

Martinez-Fernandez, G., Duval, S., Kindermann, M., Schirra, H. J., Denman, S. E. and McSweeney, C. S. 2018. 3-NOP vs. Halogenated compound: methane production, ruminal fermentation and microbial community response in forage fed cattle. *Frontiers in Microbiology* 9(1582), 1582. doi:10.3389/fmicb.2018.01582.

McAllister, T. A. and Newbold, C. J. 2008. Redirecting rumen fermentation to reduce methanogenesis. *Australian Journal of Experimental Agriculture* 48(2), 7–13. doi:10.1071/EA07218.

McAllister, T. A., Meale, S. J., Valle, E., Guan, L. L., Zhou, M., Kelly, W. J., Henderson, G., Attwood, G. T. and Janssen, P. H. 2015. RUMINANT NUTRITION SYMPOSIUM: use of genomics and transcriptomics to identify strategies to lower ruminal methanogenesis. *Journal of Animal Science* 93(4), 1431–49. doi:10.2527/jas.2014-8329.

McGinn, S. M., Beauchemin, K. A., Coates, T. and Colombatto, D. 2004. Methane emissions from beef cattle: effects of monensin, sunflower oil, enzymes, yeast, and fumaric acid. *Journal of Animal Science* 82(11), 3346–56. doi:10.2527/2004.82113346x.

Milich, L. 1999. The role of methane in global warming: where might mitigation strategies be focused? *Global Environmental Change* 9(3), 179–201. doi:10.1016/S0959-3780(98)00037-5.

Moate, P. J., Deighton, M. H., Williams, S. R. O., Pryce, J. E., Hayes, B. J., Jacobs, J. L., Eckard, R. J., Hannah, M. C. and Wales, W. J. 2016. Reducing the carbon footprint of Australian milk production by mitigation of enteric methane emissions. *Animal Production Science* 56(7), 1017–34. doi:10.1071/AN15222.

Moate, P. J., Jacobs, J. L., Hannah, M. C., Morris, G. L., Beauchemin, K. A., Alvarez Hess, P. S., Eckard, R. J., Liu, Z., Rochfort, S., Wales, W. J. and Williams, S. R. O. 2018. Adaptation responses in milk fat yield and methane emissions of dairy cows when wheat was included in their diet for 16 weeks. *Journal of Dairy Science* 101(8), 7117–32. doi:10.3168/jds.2017-14334.

Moate, P. J., Williams, S. R. O., Deighton, M. H., Hannah, M. C., Ribaux, B. E., Morris, G. L., Jacobs, J. L., Hill, J. and Wales, W. J. 2019. Effects of feeding wheat or corn and of rumen fistulation on milk production and methane emissions of dairy cows. *Animal Production Science* 59(5), 891–905. doi:10.1071/AN17433.

Nagaraja, T. G. and Chengappa, M. M. 1998. Liver abscesses in feedlot cattle: a review. *Journal of Animal Science* 76(1), 287–98. doi:10.2527/1998.761287x.

Nisbet, D. J. and Martin, S. A. 1993. Effects of fumarate, L-malate, and an *Aspergillus oryzae* fermentation extract on D-lactate utilization by the ruminal bacterium *Selenomonas ruminantium*. *Current Microbiology* 26, 136–.

Odongo, N. E., Bagg, R., Vessie, G., Dick, P., Or-Rashid, M. M., Hook, S. E., Gray, J. T., Kebreab, E., France, J. and McBride, B. W. 2007. Long-term effects of feeding monensin on methane production in lactating dairy cows. *Journal of Dairy Science* 90(4), 1781–8. doi:10.3168/jds.2006-708.

Okine, E. K., Mathison, G. W. and Hardin, R. T. 1989. Effects of changes in frequency of reticular contractions on fluid and particulate passage rates in cattle. *Journal of Animal Science* 67(12), 3388–96. doi:10.2527/jas1989.67123388x.

Opio, C., Gerber, P. and Mottet, A. 2012. *Greenhouse Gas Emission from Ruminant Supply Chains*. Food and Agriculture Organization/World Health Organization.

Ortega-Cerrilla, M. E. and Mendoza-Martínez, G. 2003. Starch digestion and glucose metabolism in the ruminant: a review. *Interciencia* 28, 380–6.

Owens, F. N., Secrist, D. S., Hill, W. J. and Gill, D. R. 1998. Acidosis in cattle: a review. *Journal of Animal Science* 76(1), 275–86. doi:10.2527/1998.761275x.

Patra, A. K. and Saxena, J. 2009. Dietary phytochemicals as rumen modifiers: a review of the effects on microbial populations. *Antonie van Leeuwenhoek* 96(4), 363–75. doi:10.1007/s10482-009-9364-1.

Pitta, D. W., Pinchak, W. E., Indugu, N., Vecchiarelli, B., Sinha, R. and Fulford, J. D. 2016. Metagenomic analysis of the rumen microbiome of steers with wheat-induced frothy bloat. *Frontiers in Microbiology* 7(689), 689. doi:10.3389/fmicb.2016.00689.

Plaizier, J. C., Krause, D. O., Gozho, G. N. and McBride, B. W. 2008. Subacute ruminal acidosis in dairy cows: the physiological causes, incidence and consequences. *The Veterinary Journal* 176(1), 21–31. doi:10.1016/j.tvjl.2007.12.016.

Rowe, J. B., Choct, M. and Pethick, D. W. 1999. Processing cereal grains for animal feeding. *Australian Journal of Agricultural Research* 50(5), 721–36. doi:10.1071/AR98163.

Russell, J. B. 2002. *Rumen Microbiology and Its Role in Ruminant Nutrition*. Cornell University Press, Ithaca, NY.

Russell, J. B. and Hespell, R. B. 1981. Microbial rumen fermentation. *Journal of Dairy Science* 64(6), 1153–69. doi:10.3168/jds.S0022-0302(81)82694-X.

Russell, J. B. and Robinson, P. H. 1984. Compositions and characteristics of strains of *Streptococcus bovis*. *Journal of Dairy Science* 67(7), 1525–31. doi:10.3168/jds.S0022-0302(84)81471-X.

Russell, J. B. and Rychlik, J. L. 2001. Factors that alter rumen microbial ecology. *Science* 292(5519), 1119–22. doi:10.1126/science.1058830.

Russell, J. B. and Strobel, H. J. 1989. Effect of ionophores on ruminal fermentation. *Applied and Environmental Microbiology* 55(1), 1–6. doi:10.1128/AEM.55.1.1-6.1989.

Sauvant, D. and Giger-Reverdin, S. 2007. Empirical modelling by meta-analysis of digestive interactions and CH_4 production in ruminants. In: *Energy and Protein Metabolism and Nutrition*. Publication – European Association for Animal Production (EAAP), 124. Presented at 2nd EAAP International Symposium on Energy and Protein Metabolism and Nutrition, Vichy, FRA, 09–13 September 2007. Wageningen Academic Publishers, Wageningen, the Netherlands, pp. 561–2. Available at: https://prodinra.inra.fr/record/13493.

Schultz, J. E. and Breznak, J. A. 1979. Cross-feeding of lactate between *Streptococcus lactis* and *Bacteroides* sp. isolated from termite hindguts. *Applied and Environmental Microbiology* 37(6), 1206–10. doi:10.1128/AEM.37.6.1206-1210.1979.

Seshadri, R., Leahy, S. C., Attwood, G. T., Teh, K. H., Lambie, S. C., Cookson, A. L., Eloe-Fadrosh, E. A., Pavlopoulos, G. A., Hadjithomas, M., Varghese, N. J., Paez-Espino, D., Hungate1000 Project Collaborators, Palevich, N., Janssen, P. H., Ronimus, R. S., Noel, S., Soni, P., Reilly, K., Atherly, T., Ziemer, C., Wright, A.-D., Ishaq, S., Cotta, M., Thompson, S., Crosley, K., McKain, N., Wallace, R. J., Flint, H. J., Martin, J. C., Forster, R. J., Gruninger, R. J., McAllister, T., Gilbert, R., Ouwerkerk, D., Klieve, A., Jassim, R. A., Denman, S., McSweeney, C., Rosewarne, C., Koike, S., Kobayashi, Y., Mitsumori, M., Shinkai, T., Cravero, S., Cucchi, M. C., Perry, R., Henderson, G., Creevey, C. J., Terrapon, N., Lapebie, P., Drula, E., Lombard, V., Rubin, E., Kyrpides, N. C., Henrissat, B., Woyke, T., Ivanova, N. N. and Kelly, W. J. 2018. Cultivation and sequencing of rumen microbiome members from the Hungate1000 Collection. *Nature Biotechnology* 36, 359.

Steinfeld, H., Gerber, P., Wassenaar, T., Castel, V., Rosales, M., Rosales, M. and de Haan, C. 2006. *Livestock's Long Shadow: Environmental Issues and Options*. Food and Agriculture Organization, Rome, Italy.

Stevens, E., Armstrong, K., Bezar, H., Griffin, W. and Hampton, J. 2004. Fodder oats an overview. In: Suttie, J. M. and Reynolds, S. G. (Eds), *Fodder Oats: a World Overview*. Plant Production and Protection Series No. 33. Food and Agriculture Organization, Rome.

Tedeschi, L. O. and Nagaraja, T. G. 2020. *Phibro Rumen Health Compendium*. Texas A&M University Press, College Station, TX.

Theurer, C. B. 1986. Grain processing effects on starch utilization by ruminants. *Journal of Animal Science* 63(5), 1649–62. doi:10.2527/jas1986.6351649x.

Thiele, J. H. and Zeikus, J. G. 1988. Control of interspecies electron flow during anaerobic digestion: significance of formate transfer versus hydrogen transfer during syntrophic methanogenesis in flocs. *Applied and Environmental Microbiology* 54(1), 20–9. doi:10.1128/AEM.54.1.20-29.1988.

Toland, P. 1976. The digestibility of wheat, barley or oat grain fed either whole or rolled at restricted levels with hay to steers. *Australian Journal of Experimental Agriculture* 16(78), 71–5. doi:10.1071/EA9760071.

Van Kessel, J. S. and Russell, J. B. 1996. The effect of pH on ruminal methanogenesis. *FEMS Microbiology Ecology* 20(4), 205–10. doi:10.1016/0168-6496(96)00030-X.

Verdu, M., Bach, A. and Devant, M. 2015. Effect of concentrate feeder design on performance, eating and animal behavior, welfare, ruminal health, and carcass quality in Holstein bulls fed high-concentrate diets. *Journal of Animal Science* 93(6), 3018–33. doi:10.2527/jas.2014-8540.

Wallace, R. J., Wood, T. A., Rowe, A., Price, J., Yanez, D. R., Williams, S. P. and Newbold, C. J. 2006. Encapsulated fumaric acid as a means of decreasing ruminal methane emissions. *International Congress Series* 1293, 148–51. doi:10.1016/j.ics.2006.02.018.

Wells, J. E., Krause, D. O., Callaway, T. R. and Russell, J. B. 1997. A bacteriocin-mediated antagonism by ruminal lactobacilli against *Streptococcus bovis*. *FEMS Microbiology Ecology* 22(3), 237–43. doi:10.1111/j.1574-6941.1997.tb00376.x.

Wilson, B. K., Holland, B. P., Step, D. L., Jacob, M. E., Vanoverbeke, D. L., Richards, C. J., Nagaraja, T. G. and Krehbiel, C. R. 2016. Feeding wet distillers grains plus solubles with and without a direct-fed microbial to determine performance, carcass characteristics, and fecal shedding of *Escherichia coli* O157:H7 in feedlot heifers. *Journal of Animal Science* 94(1), 297–305. doi:10.2527/jas.2015-9601.

Wolfe, R. S. 1993. An historical overview of methanogenesis. In: Ferry, J. G. (Ed.), *Methanogenesis*. Chapman and Hall Publishers, New York, NY, pp. 1–32.

Wolin, M. J. 1975. Interactions between the bacterial species in the rumen. In: McDonald, W. and Warner, A. C. I. (Eds), *Digestion and Metabolism in the Ruminant*. Univ. New Eng. Pub. Unit, Armidale, Australia, pp. 134–48.

Wood, T. A., Wallace, R. J., Rowe, A., Price, J., Yáñez-Ruiz, D. R., Murray, P. and Newbold, C. J. 2009. Encapsulated fumaric acid as a feed ingredient to decrease ruminal methane emissions. *Animal Feed Science and Technology* 152(1–2), 62–71. doi:10.1016/j.anifeedsci.2009.03.006.

Wright, A. G. and Klieve, A. V. 2011. Does the complexity of the rumen microbial ecology preclude methane mitigation? *Animal Feed Science and Technology* 166–167, 248–53. doi:10.1016/j.anifeedsci.2011.04.015.

Xia, Y., Kong, Y., Seviour, R., Yang, H. E., Forster, R., Vasanthan, T. and McAllister, T. 2015. In situ identification and quantification of starch-hydrolyzing bacteria attached to barley and corn grain in the rumen of cows fed barley-based diets. *FEMS Microbiology Ecology* 91(8), fiv077. doi:10.1093/femsec/fiv077.

Zhu, F. 2017. Barley starch: composition, structure, properties, and modifications. *Comprehensive Reviews in Food Science and Food Safety* 16(4), 558–79. doi:10.1111/1541-4337.12265.

Chapter 2

The use and abuse of cereals, legumes and crop residues in rations for dairy cattle

Michael Blümmel, International Livestock Research Institute (ILRI), Ethiopia; A. Muller, Research Institute of Organic Agriculture (FiBL), and ETH Zürich Switzerland; C. Schader, Research Institute of Organic Agriculture (FiBL), Switzerland; M. Herrero, Commonwealth Scientific and Industrial Research Organization, Australia; and M. R. Garg, National Dairy Development Board (NDDB), India

1 Introduction

2 Current and future levels of animal sourced food (ASF) production

3 Dairy ration compositions and current and projected feed demand and supply

4 Context specificity of feed demand and supply

5 Ration composition and ceilings to milk productivity

6 Optimizing the feed–animal interface: ration balancing in intensive and extensive dairy systems

7 Summary

8 Where to look for further information

9 References

1 Introduction

Globally, livestock contributes 40% to agricultural GDP, employs more than a billion people and creates livelihoods for more than 1 billion poor. From a nutritional standpoint, livestock contributes about 30% of the protein in human diets globally and more than 50% in developed countries. At the same time livestock contributes 18% of the total global greenhouse gas (GHG) emissions from human sources, requires 30% of land surface and 70% of agricultural land and is an important agent of land degradation, deforestation and nitrogen (N) and phosphorus (P) cause eutrophication in water bodies (Steinfeld et al., 2006). Feed sourcing and feeding is at the very interface where the 'good' and the 'bad' of livestock production are negotiated (Blümmel et al., 2010). Feed cost and feed conversion efficiencies largely determine the economic performance of livestock. Feed production takes the bulk of water invested in livestock production (Singh et al., 2004), competes with food production through allocation of arable land and restricts organic matter availability for soil health, while inefficient feed conversion contributes to emissions and environmental pollution. However, feedstuffs are not a homogenous group in how they

http://dx.doi.org/10.19103/AS.2016.0006.14

interact with natural resource use and environment but can be diversified in a number of ways. Diet formulation and animal requirements can be a starting point, classifying feeds into protein, energy and minerals/trace elements/vitamins. These in turn can be grouped into concentrates, basal diets, forages, roughages, supplements and so on with boundaries often fluent. Another way for stratification could be derived from feed sourcing. Some feed resources compete directly with human nutrition, while others do not. Among the latter count grasslands not suitable for crop production, by-products from cropping, the crop residues (straws, stover, haulms), agro by-products (brans, cakes, threshing residues) and brewery, biofuel by-products and so on. By-products are generally associated with smaller negative environmental footprints than primary produce use such as grains since inputs such as water and land are allocated across several products such as grains, bran and straw or oil, cakes and haulms (Blümmel et al., 2009). Feeds that do not directly compete with human nutrition, such as planted forages, do so indirectly by occupying arable land and consuming nutrients and water (Schader et al., 2015), and their negative environmental footprints can be severe. On the other hand, negative environmental footprints relative to a unit of animal sourced food (ASF) are generally inversely associated with the level of intensification which in turn requires quality feedstuffs (Gerber et al., 2013). Choice of feed sourcing and feeding is, therefore, multidimensional, requiring very context-specific trade-off analysis and optimization strategies. A discussion about use and abuse of cereals, legumes and crop residues in rations for dairy cattle will, therefore, have to look not only at the feed–animal interface but also at the natural resource use–feed interface. This chapter will review key elements in trade-off analysis and explore opportunities and limitations to making better use of existing feed resources and of producing more feed biomass of higher fodder quality.

2 Current and future levels of animal sourced food (ASF) production

Consumption of ASF is projected to increase substantially in the decades to come (Delgado et al., 1999; Alexandratos and Bruinsma, 2012). Intuitively, rising demand for ASF should increase feed demand and therefore aggravate the negative aspects of livestock production. However, it is important to realize that this increased demand is predicted to happen almost entirely in so-called emerging economies and developing countries where urbanization and rising incomes change food preference towards ASF (Delgado et al., 1999). For example, milk demand in developing countries in the second decade of the twenty-first century is forecast to increase by about 130 000 000 million tonnes (Gerosa and Skoet, 2012). This development is generally welcome by human nutritionist and development practitioners. First, current low ASF consumption in these countries is widely associated with malnutrition and particularly delayed child development ('stunting'). Second, increased demand for ASF provides market opportunities and income for small holder and even landless livestock keepers, thereby providing pathways out of poverty (Kristjianson, 2009; CGIAR, 2012). In contrast, for developed countries, nutritionists and health experts as well as environmentalists and animal right activists describe a situation of serious overconsumption of ASF, arguing for a strategy of contraction (reduction in ASF consumption in developed countries) and convergence (increased ASF consumption in developing and emerging countries to attain nutritionally recommended levels); see McMichael et al. (2007). Contraction in ASF consumption in the developed world would result in a reduction of feed demand, lessening the impact of feed resourcing on food security and the environment.

As outlined in the livestock revolution scenario (Delgado et al., 1999), consumption of ASF will rise in developing and emerging countries. Current per capita daily meat and milk consumption pattern in developed and developing countries and some divergent regions are summarized in Table 1 and compared with recommended levels of ASF production. Relative to recommended daily levels of animal-derived protein consumption of 20 g, overconsumption is clear in developed countries and Latin America, while average consumption in developing countries is below recommended levels (McMichael et al., 2007). When expressed in terms of animal protein assuming a protein content of 22.2% and 3.4% in meat and milk, respectively, ASF consumption in the developed world would be 3.6 times the recommended level, while the developed world as a whole would consume 80% of the recommended levels, with Africa attaining only 49% of this level.

Gerosa and Skoet (2012) analysed current and future demand for milk. During the current decade milk production in the developing world is predicted to far outpace production in the developed world by about 100 000 000 tonnes, with predicted yearly growth rates in consumption till 2050 being 2.47% and 0.6% in the developing and developed world, respectively (Table 2).

Table 1 Current daily per capita meat and milk consumption in different spheres of the globe and convergent animal protein level consumption levels recommended for 2050 based on human health and environmental considerations

Country/category	Current consumption (g/d)		Recommended consumption (g/d)
	Meat	Milk	
Developed countries	224 (50)[-1]	586 (20)	
Latin America	147 (33)	310 (11)	
			90 g/d or 20 g/d animal protein
Developing countries	47 (10)	151 (5)	
Africa	31 (7)	83 (3)	

[-1]Values in brackets indicate protein consumption; Data modified from McMichael et al. (2007) and Gerosa and Skoet (2012)

Table 2 Production of milk and projected growth and consumption rates

	Production (thousand tons)		Growth rate (%)	Consumption rates (%)
	Average 2008–10	2020	2001–20	2001–50
World	692 647	852 898	2.0	1.2
Developed countries	318 980	349 769	0.6	0.32[-1]
Developing countries	373 667	503 128	3.2	2.47

[-1]Excludes former centrally planned economies where the growth rate is negative (−0.2).
Data summarized from Tables 6 and 7 of Gerosa and Skoet (2012).

3 Dairy ration compositions and current and projected feed demand and supply

The global use of feeds for livestock between 1992 and 2000 was estimated at 4.6–5.3 billion tonnes of dry matter per year (Bouwman et al., 2005; Wirsenius et al., 2010; Herrero et al., 2013). Using the data from the model used in Schader et al. (2015), we arrive at a similar estimate of 4.7 billion tonnes DM per year for an average over the years 2005–9. Accepting that feeds are not a homogenous group, total feed biomass is not a satisfactory quantity and some sub-classification/disaggregation is warranted. Herrero et al. (2013) used four feed categories: (1) grains, usually fed as concentrates, (2) grasses from direct grazing and silage grasses, (3) stover from fibrous crop residues and (4) occasional feeds comprising cut and curry forages and legumes and roadside grasses, and disaggregated feed from these four categories according to geographical region, production system and animal species. For the present paper, we calculated global dairy rations on the basis of these four feed categories, using the supplementary data provided by Herrero et al. (2013). According to these calculations, grasses were the almost exclusive feed resources for dairy bovines in pure livestock systems and the major ration component in mixed crop livestock and urban and peri-urban systems across all agro-ecological zones (Table 3). Grains were major components in mixed crop livestock systems and in urban and peri-urban systems.

On a global scale the proportional contributions of grasses, grains, stover and occasional feeds would be 67.5%, 16.0%, 11.7% and 4.5%, respectively (Fig. 1). However, considerable differences exist between key regions of the developed and developing world in the composition of feed resources and associated dairy productivity (Table 4; these data were also calculated from the supplementary data set provided by Herrero

Table 3 Percentage grass, grain, crop residues and occasional feed contributing to global dairy rations in different systems (calculated from supplementary material of Herrero et al. 2013)

System	Grass	Grain	Stover	Occasional
LGA	95.0	0.0	3.0	3.0
LGH	88.0	5.0	0.0	7.0
LGT	96.0	3.0	1.0	0.0
Mean PLS	**93.05**	**2.7**	**1.3**	**3.3**
MRA	38.0	29.0	31.0	2.0
MRH	50.0	18.0	24.0	8.0
MRT	55.0	31.0	13.0	1.0
Mean CLS	**47.7**	**26.0**	**22.7**	**3.7**
Others	67.0	18.0	9.0	6.0
Urban	52.0	24.0	13.0	12.0
Mean Others/Urban	**59.5**	**21.0**	**11.0**	**9.0**

PLS = Pure Livestock System, CLS = Crop Livestock System
LGA, LGH, LGT = Livestock systems in arid, humid and temperate regions; MRA, MRH, MRT = mixed crop livestock systems in arid, humid and temperate regions. Data calculated from supplementary material provided by Herrero et al. (2013) and expressed on a dry matter basis.

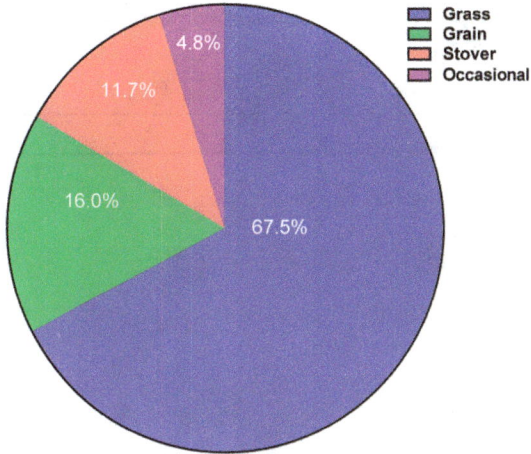

Figure 1 Global key feed resources for dairy (calculated from supplementary material provided by Herrero et al., 2013).

et al., 2013). While grasses overwhelmingly dominated feed resources in all pure livestock systems regardless of the region, composition of dairy ration in mixed crop livestock systems could vary substantially across the region with relatively high grain usage in the Middle East/North Africa and North America.

It is important to realize that the data of Herrero et al. (2013) were mostly based on secondary data sets such as cropping, livestock census and modelling of livestock populations. Also, the data sets presented in Table 3 and 4 and Fig. 1 were essentially interpreted in a relative and descriptive fashion, neglecting actual livestock numbers in the different systems. Very recently, Tricarico (2016) described average dairy rations in the United States on the basis of surveys of 350 representative dairy farms using weighted average rations for calves, open heifers, bred heifers, first calf heifers, springers and lactating and dry cows. Tricarico (2016) distinguished four feed classes: (1) forages with 9 feeds, (2) forage by-products with one feed (wheat straw), (3) concentrates with 10 feeds/feed ingredients and (4) concentrate by-products with 13 feeds/feed ingredients. Table 5 presents the average contribution of forages (wheat straw was added to forages), concentrates and concentrate by-products as synthesized from the data presented by Tricarico (2016). For each feed class the three most important feeds/feed ingredients were also calculated.

The data from Tricarico (2016) show that forages contribute on average about 53.5% to current dairy rations in the United States and concentrate components (concentrate and concentrate by-products) about 46.5%. It would appear that these data broadly agree with the estimates of dairy rations presented in Table 4 for mixed crop–livestock systems in North America. Tricarico (2016) further disaggregated dairy feed resources and reported that only eight crops accounted for 80% of the US dairy feed (corn 42%, alfalfa 22%, wheat 3.1%, soyabean 3%, canola 1.8%, sorghum 1.7%, barley 1.4% and cottonseed 1.4%). Food, fuel and fibre by-products (14 ingredients) accounted for 19% of the US dairy feeds. Using calculations of feed/feed ingredient cell contents (100-NDF), which are digestible by humans, Tricarico (2016) estimated the amount of potentially human-edible dairy feed to be at most 20% of the average US dry matter dairy ration. This proportion was reduced

Table 4 Percentage grass, grain, stover and occasional feed contributing to global dairy rations in different global regions and systems

Region/system	Grass	Grain	Stover	Occasional	Milk yield kg/d
			%		
East Asia					
PLS	96.7	3.3			2.6
CLS	21.0	13.7	61.3	4.0	3.5
OTUR	84.5	15.5			10.0
Europe					
PLS	87.0	13.0		5.0	16.1
CLS	68.7	26.7		8.5	16.8
OTUR	72.0	19.5			16.9
Latin America					
PLS	91.3	1.0		7.7	4.1
CLS	43.3	18.3	19.0	19.0	10.3
OTUR	76.0	5.5		18.5	4.1
M. East and N. Africa					
PLS	100				3.0
CLS	33.7	46.0	20.3		7.6
OTUR	55.0	45.0			6.9
North America					
PLS	NA	NA	NA	NA	NA
CLS	59.3	40.7			27.4
OTUR	70.0	30.0			25.0
Oceania					
PLS	84.0	16.0			15.2
CLS	84.0	16.0			14.9
OTUR	44.5	10.5		45.0	10.3
South Asia					
PLS	100				3.1
CLS	31.3	36.3	31.7		5.6
OTUR	7.0	37.0	56.5		2.9
South East Asia					
PLS	100				1.7
CLS	29.3	15.3	55.00		3.0
OTUR	80.0		20.00		1.4
Sub-Saharan Africa					
PLS	89.3	4.3	5.0	1.3	2.4
CLS	54.7	10.7	29.7	4.7	1.9
OTUR	89.5	2.0	5.5	3.0	3.1

(PLS = Pure Livestock System; CLS = Crop Livestock System, OTUR = Others and Urban). Data expressed on a dry matter basis.

to 2.2% when actual demand as human edible food/human edible food ingredients by the food industry was included; cell contents might be digestible by human enzymes but not conform to food preferences and thus will have no such market (Tricarico, 2016). These findings appear to suggest that current dairy feeding practices in the United States do not necessitate revisiting of feed sourcing and feeding practices as far as nutrient competition between humans and dairy animals is concerned.

However, this point of view neglects the high proportion of corn silage (22.4%), alfalfa silage (11.5%) and alfalfa hay (10.8%) in the dry matter of the average US dairy ration, all of which will need good arable land, and in the case of alfalfa, considerable amounts of water. These resources are not available for direct human food production. This way of looking at feed resources was presented by Schader et al. (2015) who distinguished between feed resources that compete with direct human food crop production (food competing feedstuffs, FCF) and those that do not, which were defined as grassland assumed to be not suitable as arable land and by-products. Table 6 lists regional dairy feed resources classified according to Schader et al. (2015). Compared to the classification derived from

Table 5 Average national US dairy dry matter rations extrapolated from surveys of 350 representative dairy farms. Summarized into forages, concentrates and concentrate by-products from 33 individual feed ingredients reported by Tricarico (2016)

Feed class/three highest sub-classes	Average contribution to dairy ration (%)
Forages	53.5
Silage	38.6
Hay	13.5
Pasture	1.4
Concentrates	28.9
Corn grain	9.7
Miscellaneous	5.0
High moisture corn grain	2.8
Concentrate by-products	17.5
Distillers grain	4.1
Condensed gluten feed	2.6
Cotton seed	2.2

Table 6 Shares of grassland, food competing feedstuffs (FCF) and by-products in dairy rations in regions of the world according to the classification of Schader et al. 2015

Region	Grass	FCF (%)	By-products/residues (%)
East Asia	55	18	27
Europe	71	26	3
Latin America	62	24	14
M. East and N. Africa	50	27	23
North America	61	36	3
Oceania	71	28	1
South Asia	29	22	49
South East Asia	62	13	25
Sub-Saharan Africa	69	9	21

Herrero et al. (2013), grassland contributed about 10% units less to dairy rations, though remaining the most important feed resource, except for South Asia.

Key global dairy feed resources such reclassified are presented in Fig. 2. This reclassification suggests that about 23% of global dairy rations consist of FCF, grasslands not suitable for crop cultivation contribute about 59% and by-products and crop residues contribute about 19% (compare Fig. 1 and 2). However, application of Schader et al. (2015) classifications to Tricarico (2016) US dairy feed survey challenges this perception about rather moderate competitiveness of dairy feed with direct human food production, since then only 36.4% of the average US dairy ration would consist of none food competing feedstuffs (Fig. 3). The Schader et al. (2105) estimate for North Africa (Table 6) proposes

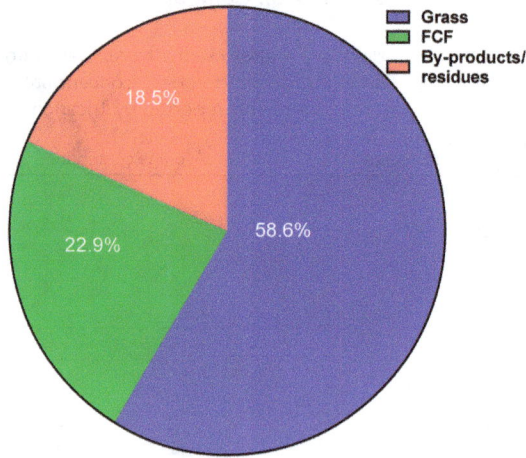

Figure 2 Global key feed resources for dairy according to the classification of Schader et al. (2015) into grassland, food competing feedstuffs (FCF) and by-products.

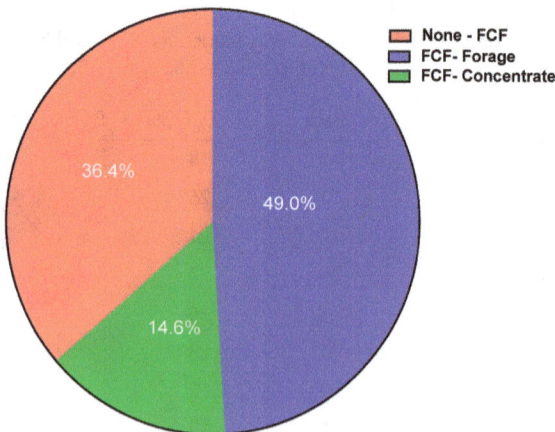

Figure 3 Average US dairy rations from Tricarico (2016) reclassified based on Schader et al. (2015).

this feed fraction to be more than 63%. This disagreement highlights the importance of ground-truthing modelled feed resources by surveys or at the very least by triangulations using independent data sets.

Whatever the reason for the above disagreements on average dairy rations, feed resourcing that does not include, or that reduces, the feeding of FCF will affect the production of ASF. To assess these effects, Schader et al. (2015) generated two scenarios, one based on FAO projections on food demand and production in 2050 (Alexandratos and Bruinsma, 2012), while the other on a gradual reduction in the use of FCF down to 0%. Table 7 summarizes the effect of complete elimination of FCF from livestock diets relative to the baseline (mean values calculated across the period 2005–9) and the FAO scenario.

Elimination of FCF from livestock rations by 2050 would have major effects on a number of monogastric animals and per capita consumption of energy and protein from ASF, the consumption of which would fall below recommended levels (compare Tables 1 and 7). While reduction in the availability of ASF per capita is strongest for meat, milk availability would also decrease by 43% (Fig. 4). This, of course, would conflict with the increased demand for milk in exactly those areas, that is, the tropic and subtropics (Table 2).

Table 7 Effect of eliminating food competing feedstuffs (FCF) from the diet on livestock numbers (billions) and supply of animal food energy (kcal/capita/day) and protein (g/capita/day)

Item	Baseline (2005–9)	Reference scenario (2050)	Zero FCF (2050)
Large ruminants (cattle + buffalo)	1.57	2.12	1.71
Small ruminants (sheep and goat)	1.96	2.99	2.52
Pigs	0.92	1.17	0.11
Chicken	17.56	33.85	5.19
Supply of animal food energy	414	515	151
Supply of animal protein	26.2	31.2	8.6

Calculated from Schader et al. (2015)

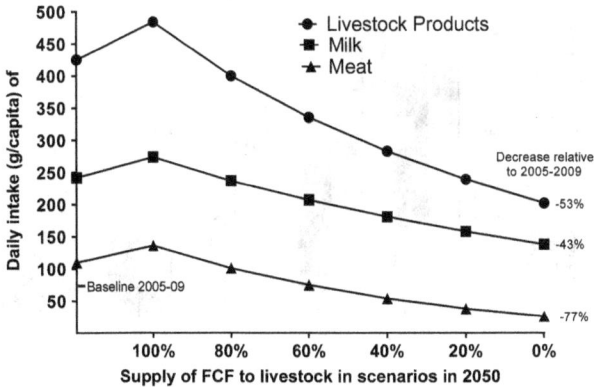

Figure 4 Per capita availability of animal sourced food (ASF) in 2050 with stepwise elimination of food competing feed (FCF) from livestock rations. Source: Adapted from Table 2 of Schader et al., 2015.

Country-specific studies on the potential of grassland and by-products-based animal production, better capture local characteristics than such global assessments, for example, regarding domestic feed supply (by-product availability and grassland productivity) or animal varieties and feed conversion aspects. Van Kernebeek et al. (2015), for example, investigated this for the Netherlands and found that a reduction of animal protein in total protein supply down to 12% would result in lowest land use. Lower shares of animal products lead to increased land use as by-products and grasslands could not be used optimally anymore. Addressing a number of grassland-based production scenarios in Sweden, Röös et al. (2016) reached similar conclusions. These studies demonstrate the classification of feedstuffs according to the degree of competition to food production to be particularly relevant for optimizing dairy systems with regard to their environmental and economic performance especially because the use of by-products is considered to create less incentives for production and because it can substitute a share of FCF commonly used. Therefore, lower proportions of environmental impacts are allocated to these products, even more so if economic allocation instead of mass allocation is applied in environmental assessments (IES, 2010).

4 Context specificity of feed demand and supply

A number of studies have projected feed use growth rates between 2.9% and 3.3% accumulating to 6.5 to 8 billion tonnes of dry matter per year in 2030 (Bouwman et al., 2005; Wirsenius et al., 2010; Havlik et al., 2014). Most of this growth is expected in the tropics and subtropics, the areas that exhibit the highest increase in the demand for livestock products. It is important to realize that productivity in these countries is very low (see also Table 4). For example in India, the biggest milk producer in the world, average across herd (indigenous cattle, cross-bed cattle and buffalo) milk yield in 2005–6

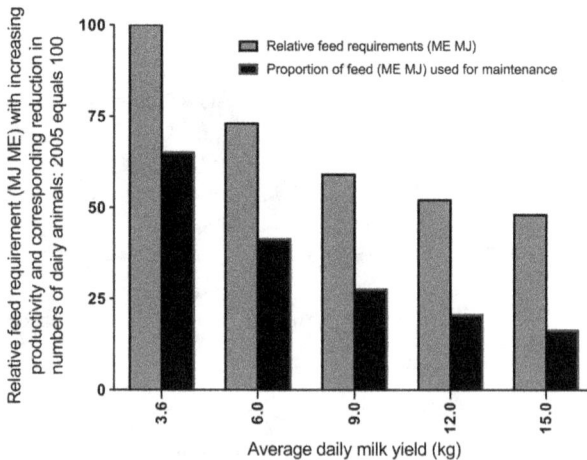

Figure 5 Feed requirement in dependency of per dairy animal productivity: the Indian scenario. Source: From Blummel et al., 2013.

was 3.61 kg per day (Blümmel et al., 2013a). In 2005–6, India produced about 82 million tonnes of milk from a dairy livestock population of 69 759 000. If average daily milk yield could be increased to 15 kg per day with concomitant reduction in numbers of dairy animals, these 82 million tonnes could theoretically be produced with about half the feed, with the proportion of feed used for animal maintenance requirement decreasing from 66% at yields of 3.6 kg to 34% at yields of 15 kg, see Fig. 5 (data calculated from Blümmel et al., 2013a). In other words feed demand for the production of a given unit of milk varies hugely in dependency of per animal productivity, or of the level of intensification of dairy production.

It is important to note that productivity levels of 15 kg of milk daily are clearly within our reach, even on diets consisting of more than 90% of by-products, that is, non-FCFs. Anandan et al. (2010), in collaboration with the private feed manufacturers Miracle Fodder and Feeds PVT LTD in India (Shah, 2007), designed and tested so-called densified total mixed ration (DTMR) feed blocks that consisted largely of by-products such as sorghum stover (50%), cereal bran (9% rice and wheat), legume husks and hulls (9% pigeon pea, chickpea and mung beans), oil cakes (18%) with the rest contributed by sugar cane molasses (8%), maize grain (4%) and a mix of urea, minerals and vitamins (2%). In a series of trials with commercial dairy buffalo producers, Anandan et al. (2010) varied the quality of the sorghum stover in the DTMR by purchasing and incorporating low cost (in vitro organic matter digestibility (IVOMD = 47%)) and premium sorghum stover (IVOMD = 52%) into the DTMR; the sorghum stover came from two different cultivars, marketed at fodder markets. Differences in stover quality translated into significant differences in milk production. The potential daily milk production was about 5.6 kg higher per buffalo (15.5 vs 9.9 kg) in the group fed with the complete rations based on the higher-quality sorghum stover (Blümmel et al., 2013a). This increased milk potential was due to the additive effect of higher energy content of the diet with the superior sorghum stover, and higher feed intake (Table 8).

Table 8 Milk potential in Indian dairy buffalo fed with two densified total mixed ration (DTMR; in the form of feed blocks) based on premium (52% digestibility) and low-quality (47% digestibility) sorghum stover, with total by-product proportion of feed blocks greater than 90%

	Block low-quality stover	Block premium-quality stover
Protein (%)	17.1	17.2
Metabolizable energy (MJ/kg)	7.37	8.46
Voluntary intake of feed block (kg/d)	18.0	19.7
Voluntary intake of feed block (%/kg LW[1])	3.6	3.8
ME intake (MJ/d)	132.7	166.7
Milk fat (%)	7.4	7.6
Milk potential (kg/d)	9.9	15.5
Milk potential cattle (kg/d)	14.0	21.0

Data calculated from Anandan et al. (2010) and Blümmel et al. (2013) based on the actual milk fat contents of buffalo milk. Note: milk fat in cattle was assumed to be 4% with a cross-energy content of 3.13 MJ/kg; [1]LW of buffalo was calculated by body measurements and estimated to be on average 506 and 525 kg in the low-quality and premium-quality feed block, respectively

Extrapolating the potential milk yields to dairy cattle, which have lower milk fat contents than buffalo, would result in an estimated 14 and 21 kg of milk daily in the lower- and higher-quality DTMR, respectively.

Three things are noteworthy here, and these have far-reaching implications for feeding by-products. First, a well-complementing mix of by-products such as cereal stover with leguminous threshing residues, brans and oil cakes can result in levels of productivity about thrice the current one (compare 3.61 kg/d with 9.9 (buffalo) and 14 (cattle) kg, Table 8) even with a basal diet of modest quality (IVOMD = 47%). This kind of improvement would already have very moderating effects on the relationships proposed in Fig. 4. Second, improvement in the basal diet of 5% units in IVOMD increased daily milk potential by 5.6 and 7 kg in buffalo and cattle, respectively, further moderating the relationships proposed in Fig. 4. Extensive work in the past decade between the International Livestock Research Institute (ILRI) and international and national crop improvement institutes targeting most key cereal and legume crops has shown that straws, stovers and haulms can be improved by 5–10% units in IVOMD through multidimensional crop improvement, that is, improvement of crop residue fodder quality at source through selection and conventional and molecular breeding (Sharma et al., 2010; Blümmel et al., 2013b, 2016a). Thus, sorghum and maize stover cultivars are now available with IVOMDs of 56%. When those stovers will be used to produce DTMR of the kind described above, it is unlikely that intake will be less than the one observed with a basal diet consisting of a cereal crop residue with an IVOMD of 52% (Table 8). Put differently, the milk potential on mostly by-product-based rations could be considerably higher than observed in Table 8.

Similarly, the fodder quality of legume residues after grain harvest such as haulms can vary among cultivars substantially. For example, live weight gain of sheep fed exclusively on groundnut haulms varied – cultivar dependent – between 67 and 132 g/d (Blümmel et al., 2016a). The point to be noted here is that the opportunities that lie in improving the fodder quality (and quantity for that matter) of crop residues at source through consequent phenotyping for cultivar differences and targeted further improvement through breeding and selection are currently probably underappreciated, even though this could have considerable effects on the relationships reported in Fig. 4, 5 and Table 11. Finally, related to the previous points, for a ration consisting of more than 90% of by-products, voluntary feed intakes were very high, reaching 3.6% and 3.8% of the body weights. The overall impact of improved basal diets, such as increasing IVOMD of a stover from 52 to 56%, will ultimately rest on the question whether dairy animals will respond with increasing DMI beyond the 3.8% of BW reported in Table 8 to reap the additive benefits of higher energy content of the ration with superior basal diet and higher DMI (see also below).

5 Ration composition and ceilings to milk productivity

While very considerable levels of milk production can be realized on almost completely by-product-based rations, voluntary feed intake will impose ceiling levels to such feeding regimes. Recently, Flachowsky and Meyer (2015) proposed a relationship between daily milk production, dry matter intake (DMI) and dietary roughage to concentrate ratio that suggest that at high daily milk yields of 40 kg, concentrates will have to constitute 50% of the diet (Table 9). It can be calculated that the milk yields of 10, 20 and 40 kg were associated with DMI corresponding to 1.85%, 2.46% and 3.85% of bodyweight and ME contents of diets of 9.49, 9.72 and 10.04 MJ/kg, respectively (Table 9). Intakes of 1.85% and 2.46% of bodyweight are certainly below the voluntary feed intake capacity of dairy cattle, while ME contents of 9.49% and 9.72% are suggestive of a planted forage of reasonably good quality. Expressed differently, daily milk yields of 10 to 20 kg could be achieved with forage and/or roughage components of lower fodder quality such as cereal straws and stovers.

While Flachowsky and Meyer (2015) proposed a roughage to concentrate ratio of 50:50 at daily milk yields of 40 kg, Chase and Grant (2013) reported milk yields of 35–48 kg on rations containing more than 60% of forages; the ration with the highest proportion of forage of 75% still produced 35 kg/d. 'Roughage' and 'forage' of course are somewhat imprecise definitions and Mertens (1988) proposed for cows a maximum intake of neutral detergent fibre (NDF) – an estimate of plant cell wall – of 1.2% of bodyweight. This proposition explicitly suggests a physical component of DMI regulation, as proposed in Fig. 6. While ultimate

Table 9 Ratio of dietary roughage to concentrate ratio dependency of daily milk yield proposed in literature and underlying supposition in terms of intake relative to bodyweight, net energy requirement and its partitioning between maintenance and production and calculated net (NE) and metabolizable energy (ME) content of the diet

	Milk[-1]	DMI[-1]	R:C[-1]	DMI/LW	MJ NE[-2]	MJ NE[-2]	MJ NE	NE	ME[-3]
	kg/d		%	%	Maintenance	Milk	Total	MJ/kg feed	
Dairy cow of 650 kg	10	12	90:10	1.85	38.6	31.4	70.6	5.88	9.49
	20	16	75:25	2.46	38.6	62.8	101.4	6.33	9.72
	40	25	50:50	3.85	38.6	125.6	164.2	6.57	10.04

[-1] Data adopted from Flachowsky and Meyer (2015); [-2] Calculated according to McDonald et al. (1988); [-3] Estimated by regression analysis and partitioning of NE for maintenance and production.

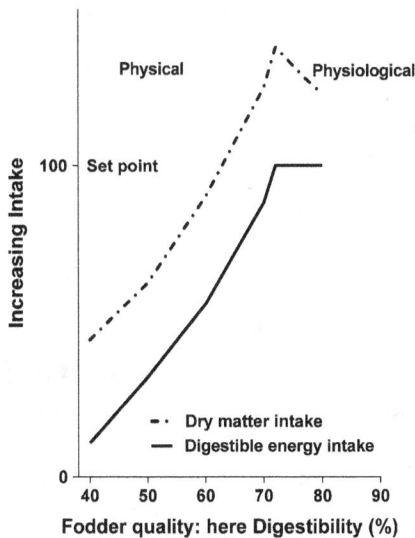

Figure 6 Relationship between fodder quality and voluntary feed intake. Source: From Van Soest, 1994.

Table 10 Summary of bodyweights (BW), neutral detergent fibre (NDF) content, dry matter intake (DMI) relative to BW, and NDF intake (NDFI) relative to BW in 80 dairy calculated from data reported by Galyean and Abney (2006)

Variable	Mean	Minimum	Maximum	SE
Bodyweight (kg)	619	527	709	5.3
NDF (%)	32.7	22.5	45.8	0.6
DMI (%/BW)	3.49	2.70	4.31	0.04
NDFI (%/BW)	1.14	0.69	1.76	0.02

causes and effects of regulation of DMI of ruminants are still contested (see Ketelaars and Tolkamp, 1992), ample empirical evidence exists for increasing DMI with increasing feed quality. While this relationship is generally well known, it is worthwhile recalling some of its key implications. First, ruminants cannot compensate for low diet quality by increasing intake. Second, increasing diet quality has two accumulating positive effects, increased energy concentration per unit feed and higher feed intake. Third, high per-animal milk productivity requires high diet quality. Fourth, voluntary feed intake is a key variable at the feed–animal interface, and will define the extent to which current feed resources can be allocated to fewer but higher-producing animals. In other words, DMI will largely determine the degree of intensification achievable on current feed resources.

Neutral detergent fibre intakes of 1.3–1.7% of bodyweight (BW) have been reported in a review of pasture research trials (Vazques and Smith, 2000) and Bargo et al. (2002) recorded NDF intakes ranging from 1.4 to 1.8% of BW in cows fed with mixed grass pastures. In a lactation trial comparing alfalfa and grass silage, Chase (2012) observed intakes of 1.6% of BW. Galyean and Abney (2006) reviewed and re-examined 80 dairy trials and calculated a linear relationship between NDF content of the diet and DMI relative to bodyweight (DMI/BW) of $y = 4.64 - 0.037$ NDF, which accounted for 62% of the variation in DMI. However, while NDF clearly restricts intake (as do other factors such as grazing time and sward characteristics), it can be summarized from the data set reviewed by Galyean and Abney (2006) that ranges in NDF intakes are indeed considerable (Table 10). Limiting putative forage/roughage intake in the design of dairy cattle rations to 1.2% of NDF relative to BW is unwarranted and maximum NDF intake is probably closer to 1.8% of BW (Table 10). These findings also agree with our previously reported work with dairy buffalos fed with by-products (Table 8), where NDF intake would be about 1.8% of BW since the DTMR contained about 48–50% of NDF.

6 Optimizing the feed–animal interface: ration balancing in intensive and extensive dairy systems

The economics of dairy feeding and the environmental footprint of dairy production demand an optimal fit between ration composition and dairy animal requirements. Total mixed rations (TMR) or feed blocks described in Table 8 probably guarantee the integrity of ration composition well, offering all ration components at the same time and minimizing feed selection. However, TMR also tend to reduce the flexibility of nutrient adjustments and many dairy farms in the United States use only one TMR for all lactating cows, despite differences in nutritional requirements of dairy cows in different lactation stages and with different body condition scores (Allen, 2009; Contreras-Govera et al., 2015). These TMRs are usually formulated to meet the requirements of the highest-producing dairy cows, resulting in the overfeeding of the lower-producing ones (Cabrera and Kalantari, 2016). Cabrera (2016) investigated nutritional grouping in five commercial dairy herds in Wisconsin and reported that having more than one nutritional group resulted on average in increased income over feed cost per cow and year of 39 US $ for two nutritional groups and of 46 US $ for three nutritional groups. The higher income was due to increased milk

Table 11 Summary of feed supply and demand in India drawn from information published by NIANP (2012) and Blümmel et al. (2014)

Feed resources	Million tons	Percentage
Greens		
From forest area	89.37	4.5
From fallow lands	23.21	1.2
From permanent pastures and grazing areas	28.70	1.4
From cultivable waste lands and miscellaneous tree crops	17.51	0.9
From planted fodder crops	303.26	15.1
		23.1
Crop residues		
Coarse straw	154.83	27.8
Fine straw	194.11	34.8
Leguminous straw	44.44	8.0
		70.6
Concentrates		
Oil cakes	15.76	2.8
Brans	13.29	2.1
Grains for feeding livestock	5.74	1.0
Chunnis (leguminous husks, hulls, etc.)	0.53	0.1
		6.3
Feed/nutrient requirements versus feed availability[1]	**Deficit (%)**	
Dry matter	6.0	
Digestible crude protein	61.0	
Total digestible nutrients	50.0	

Deficits are estimated relative to yield potential of well-managed dairy herds that is feed requirement to close yield gaps (see also Fig. 7). The yield differences between the 10% best farmers and the rest is mainly due to lack of feed.

sales and reduced feed costs, the latter mainly due to reduced requirements of rumen undegradable protein (Cabrera, 2016).

Ration balancing presents a greater challenge in more extensive dairy systems such as in India, which is still the largest dairy producer. For one, feed resources are generally more constrained and in India crop residues present the single most important feed resource, while concentrate availability is very low, contributing only 6.3% (Table 11). This obviously will pose limitations on the approaches proposed in Table 8. To remind us, the DTMR feed blocks tested consisted of 50% sorghum stover, 9% cereal bran, 9% legume husks and hulls and 18% oil cakes with the rest contributed by sugar cane molasses (8%), maize grain (4%) and a mix of urea, minerals and vitamins (2%). In other words, at least 40% of the DTMR consisted of concentrates as defined in Table 11. Production of such DTMR will be hampered by the low availability of concentrates. On the bright side though, at the same time more than 44 million tonnes of legume haulms are available (Table 11), which often provide excellent fodder on a par (and quite often superior to) with some of the concentrates such as bran and poorer-quality oil cakes, and could substitute for them (Ayantunde et al., 2014; Blümmel et al., 2016a).

Feed sourcing and feeding in extensive systems is often very opportunistic, making consistent ration design and ration offering difficult. However, as shown by the National Dairy Development Board (NDDB) of India, three key elements of ration balancing and ration design – matching protein and energy content of the ration with actual milk production, economic feed substitution and the importance of mineral supplementation – can be also implemented in these more extensive systems. The NDDB has developed a user-friendly ration balancing programme (RBP) software for preparing least cost-balanced rations, using locally available feed resources and area-specific mineral mixtures. During the past three years, NDDB has been implementing RBP in 18 major milk-producing states of India. It is envisaged to implement RBP in 40 000 villages, covering about 2.7 million dairy animals, by the year 2018–19. So far data generated by NDDB suggest that the implementation of RBP under field conditions improved ($P < 0.05$) daily milk yield in buffalo and cattle between 2 and 14% (average 3.7%) and milk fat level between 0.2 and 15% (average 4.6%) while at the same time reducing feed costs. The overall outcome was an increase in net daily income of farmers per animal ranging from 6 to 60% with an average of 16.0% (Garg et al., 2013). These findings suggest that the major beneficial effect of RBP was on lowering feed costs, rather than increasing milk production. A reason for this could reside with limited genetic potential of dairy animal in India, which if correct, would make many of the previously suggested feed intervention premature. Yet, it is unlikely that dairy animal genetics was the primary constraint to increasing milk productivity. This statement was derived from revisiting the Village Dynamic Studies in South Asia (VDSA, 2013; Blümmel et al., 2016b) of the International Crops Research Institute for the Semi-Arid Tropics (ICRISAT). This study stratified dairy farmers in India into two categories: the 10% most productive farmers and the remaining 90%. The study comprised 584 farms with 2424 local cattle and 901 improved cross-bred cattle, usually crosses between local and Holstein or Jersey cattle (the exact percentage of improved blood was not determined). Using comparable dairy animals, that is, either local cattle or cross-bred cattle, the farmers in the upper 10% category had milk yields several folds higher than the remaining 90% of the farmers (see Fig. 7). For example, in dairy farmers using local cattle the difference between the two farmers' groups was 869 kg per cow yearly. In dairy farmers using cross-bred cattle the difference between the two groups was 2309 kg per cow yearly. These are tremendous yield gaps, most certainly caused by management deficiencies

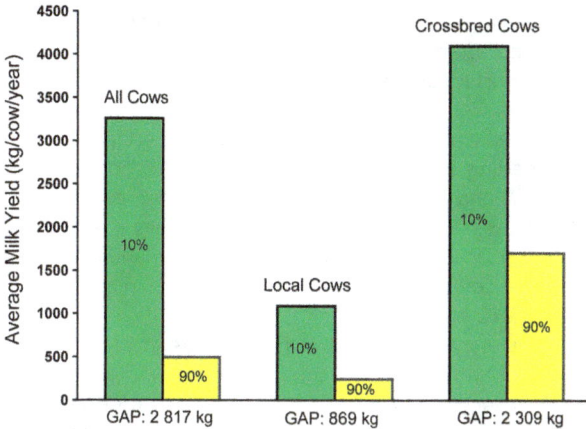

Figure 7 Yield differences in milk production between the 10% most productive farmers and the remaining 90% in India when managing comparable dairy genetics. Source: From VDSA-India, 2013 and Blummel et al., 2016b.

rather than dairy genetics, with feed quality constraints featuring foremost among the bio-physical constraints (Table 11). (There are other important constraints contributing to these yield gaps, such as risk management, market access and labour availability, but these are beyond the scope of this chapter). The findings presented in Fig. 7 also suggest that the RBP initiative of NDDB needs to move beyond the 'matching current production paradigm' to challenging dairy animals to better understand their true genetic limitations.

Closing those yield gaps by improvement of basal diets, optimization of by-product-based feeding systems, ration balancing and reallocation of feed resources to support intensification will go a long way in meeting the increased demand for milk without aggravating the negative effects of dairy production on natural resource use, food security and GHG emissions. Present life cycle assessments (LCA) show that methane (CH_4) from enteric fermentation is the largest contributor to the total GHG emissions in the smallholder dairy production system of India (Gerber et al., 2013). Therefore, improving feed conversion efficiency (FCE) for milk production with available feed resources is of paramount importance, as this will decrease the proportion of feedstuffs expended for maintenance of the animals and therefore increase both the economic and the environmental efficiency of dairy production systems. NDDB evaluated the effect of RBP on FCE in cows (n = 7090) and buffaloes (n = 4534) under field conditions and found that ration balancing improved ($P < 0.01$) FCE (kg fat corrected milk/kg DMI) from 0.61 to 0.74, 0.79 to 0.90 and from 0.80 to 0.91 in indigenous cows, cross-bred cows and buffaloes, respectively (Garg and Sherasia, 2015). Thus, through RBP, it is possible to increase FCE for milk production in cows and buffaloes, that is, to produce more milk per kg dry matter intake. Large-scale implementation of the RBP has resulted not only in higher net daily income of farmers but also in reduced carbon footprint of dairy production.

To quantify the impact of RBP on methane CH_4 emissions, NDDB has undertaken methane emission measurement studies in different agro-climatic regions of the country, using sulphur hexafluoride tracer technique. These studies showed that balanced feeding reduces CH_4 emissions (g/kg milk yield) by 17.3% ($P < 0.05$) and 19.5% ($P < 0.01$) in lactating cows and buffaloes, respectively (Garg et al., 2014). Balancing of protein, energy and minerals shifted the rumen fermentation pattern towards higher microbial efficiency, capturing feed carbon for microbial biomass production rather than short chain fatty acid production and thereby reducing CH_4 emissions. Per unit feed truly degraded in the rumen CH_4 emissions are inversely related to the efficiency of microbial production. A cradle-to-farm-gate LCA study indicated that after feeding a balanced ration, average carbon footprint of milk reduced ($P < 0.01$) from 1.6 to 1.1 and 2.3 to 1.5 kg CO_2-eq/kg fat and protein corrected milk in cows (n = 163 540) and buffaloes (n = 163 550), respectively (Garg et al., 2016). Emissions of CH_4 from enteric fermentation and CH_4 and nitrous oxide from manure management contributed 69.9%, 6.3% and 9.6% in cows; and 71.6%, 7.4% and 12.6% in buffaloes, respectively, to the baseline lifetime total GHG emissions.

Feed conversion efficiency has been shown to be one of the most important drivers for optimizing dairy systems with respect to their economic and environmental performance. At the same time it is necessary to consider potential trade-offs with other production systems as well as food demand and nutrition by considering food–feed competition. The nexus between dairy production and human food consumption is only rarely taken into account due to the complexity involved and due to the approaches that are necessary for such an assessment requiring both agronomic and economic components.

7 Summary

Feed sourcing and feeding are at the very interface where positive and negative effects from livestock are negotiated. By-product-based feed sourcing and grazing and grass production from areas not directly suitable for direct human food production reduce the food competing properties of current feeding systems. Still, foregoing the use of FCF in livestock production completely would probably necessitate lower consumption of ASF, particularly from monogastrics. It is important to realize that feed demand is very context specific and the amount of feed required to produce a unit of milk or meat can vary hugely depending on the level of intensification of the production system. In most developing and emerging countries, where most of increased demand for animal sourced food is arising, the level of intensification is very low, and often most of the feed is used for maintenance rather than productive purposes. For example, the annual dairy production of India in 2005–6 could be produced with about half the feed (in terms of ME) if average daily milk production would increase from 3.6 to 15 kg with a concomitant reduction in numbers of dairy animals. To check and reduce the negative effects of livestock and increase its positive effects, a combination of approaches is required: (a) a reduction of consumption of ASF in the Western world, (b) an increase in consumption of ASF in emerging and developing countries only to recommended levels, (c) a focus feed sourcing on by-products and other non-competing feedstuffs and (d) improving ration design and ration balancing.

8 Where to look for further information

Two of the papers cited in this work contain extensive supplementary information (Schader, C. et al. 2015): Model code and data of the SOL-Model 1.0. Available at: ftp://paper.fibl.ch (username: Paper; password: +pApler-2) as .gms and .gdx files and Herrero et al. at www.pnas.org/lookup/suppl/doi:10.1073/pnas.1308149110/-/ DCSupplemental). The website of the Livestock and Fish CRP (http://livestockfish.cgiar. org) has a wealth of information on dairy value chains in developing countries and feed resourcing and feeding problems associated with. For opportunities and limitations of food–feed/dual-purpose crops two special journal issues provide further details ('Food Feed Crops' of *Animal Nutrition and Feed Technology* in 2010 Volume 10S and 'Dual Purpose Maize' of *Field Crops Research* in 2013 Volume 153).

9 References

Alexandratos, N. and Bruinsma, J. (2012). World agriculture towards 2030/2050. The 2012 revision. In ESA Working Paper No 12-03, Agricultural Development Economics Division. Rome, Italy, FAO.

Allen, M. (2009). Grouping to increase milk yield and decrease feed costs. *Proceedings of the 20th Annual Tri-State Dairy Nutrition Conference*. The Ohio State University.

Anandan, S., Khan, A. A., Ravi, D., Jeethander Reddy and Blümmel, M. (2010). A comparison of sorghum stover based complete feed blocks with a conventional feeding practice in a Peri Urban dairy *Anim. Nutri. Feed Technol*. 10S: 23–8.

Ayantunde, A. A., Blümmel, M., Grings, E. and Duncan, A. J. (2014). Price and quality of livestock feeds in suburban markets of West Africa's Sahel: Case study from Bamako, Mali. *REMVPT* 67(1): 13–21.

Bargo, F., Muller, L. D., Delahoy, J. E. and Cassidy, T. W. (2002). Milk response to concentrate supplementation in high producing cows grazing a two pasture allowances. *J. Dairy Sci.* 85: 1777–92.

Blümmel, M., Duncan, A., Poole, J., Gerard, B. and Valbuena, D. (2010). Livestock: the good and the bad, 16–21. In Pattanaik, A. K., Verma, A. K. and Dutta, N. (Eds), *Animal Nutrition Strategies for Environmental Protection and Poverty Alleviation*, Volume 1. Proceedings of the VII Biennial Animal Nutrition Association Conference, 17–19 December 2010, Bhubaneswar, India, p. 112.

Blümmel, M., Grings, E. E. and Erenstein, O. (2013b.). Potential for dual-purpose maize varieties to meet changing maize demands: Synthesis. *Field Crops Res*. 153: 107–12.

Blümmel, M., Haileslassie, A., Samireddypalle, A., Herrero, M., Ramana Reddy, Y. and Mayberry, D. (2016b). Feed supply-demand databases as decision making tools for prioritizing livestock interventions to close yield gaps and reduce negative environmental foot prints. In Neelam Kewalramani, Nitin Tyag, Goutam Mondal, Sachin Kumar, Chander Datt, S. S. Kundu and S. K. Tomar (Eds), *Innovative Approaches for Animal Feeding and Nutritional Research*, pp. 70–81. Invited Papers of XVI Biennial Animal Nutrition Conference, 6–8 February 2016, Karnal, India, p. 372.

Blümmel, M., Haileslassie, A., Samireddypalle, A., Vadez, V. and Notenbaert, A. (2014). Livestock water productivity: feed resourcing, feeding and coupled feed-water resource data bases. *Anim. Prod. Sci.* 54 (10): 1584–93.

Blümmel, M., Homann-Kee Tui, S., Valbuena, D., Duncan, A. and Herrero, M. (2013a). Biomass in crop-livestock systems in the context of the livestock revolution. *Secheresse* 24: 330–9.

Blümmel, M., Samad, M., Singh, O. P. and Amede, T. (2009). Opportunities and limitations of food-feed crops for livestock feeding and implications for livestock-water productivity. *Rangeland J*. 31: 207–13.

Blümmel, M., Wamatu, J., Risckkowsky, B. and Moyo, S. (2016a). Opportunities and limitations of multidimensional crop improvement in grain legumes to support increased productivity in mixed cop livestock systems. International Conference on Pulses, Marrakesh, Morocco, 18–20 April 2016, Conference and Abstract Book, p. 16.

Bouwman, A. F., van der Hoek, K. W., Eickhout, B. and Soenario, I. (2005). Exploring changes in world ruminant production systems. *Agric. Syst.* 84: 121–53.

Cabrera, V. E. (2016). Impact of nutritional grouping on the economics of dairy production efficiency. *Proceedings of the Tri-State Dairy Nutrition Conference Fr Wayne*, Indiana, USA, 18–20 April 2016.

Cabrera, V. E. and Kalantari, A. S. (2016). Economics of production efficiency: Nutritional grouping. *J. Dairy Sci.* 99: 825–41.

CGIAR (2012). CGIAR Research Program on Livestock and Fish more meat and milk for and by the poor. https://livestockfish.cgiar.org.

Chase, L. E. (2012). Using grass forages in dairy rations. *Proc. Tri-State Dairy Nutr. Conf.*, pp. 75–85.

Chase, L. E. and Grant, R. J. (2013). High forage rations – what do we know. *Feedinfo Scientific Review* 15 November 2013.

Contreras-Govea, F. E., Cabrera, V. E., Armentano, L. E., Shaver, D. E., Crump, P. M., Beede, D. K. and VandeHaar, M. J. (2015). Constraints for nutritional grouping in Wisconsin and Michigan dairy farms. *J. Dairy Sci.* 98: 1336–44.

Delgado, C., Rosegrant, M., Steinfeld, H., Ehui, S. and Courbois, C. (1999). Livestock to 2020: The next food revolution. International Food Policy Research Institute, Food and Agriculture Organization of the United Nations, and International Livestock Research Institute. IFPRI Food, Agriculture and the Evironment Discussion Paper 28, Washington, DC, p. 72.

Flachowsky, G. and Meyer, U. (2015). Sustainable production of protein of animal origin – the state of knowledge. Part 1: Resources and emissions as factors affecting sustainability. *J. Anim. Feed Sci.* 24: 273–82.

Galyean, M. L. and Abney, C. S. (2006). Assessing roughage value in diets of high producing cattle. *21st Annual Southwest Nutrition and Management Conference*, 23–24 February, Tempe Arizona.

Garg, M. R. and Sherasia, P. L. (2015). Ration balancing: A practical approach for reducing methanogenesis in tropical feeding systems. In Veerasamy, S., John, G., Lance, B., Prasad, C. S. (Eds), *Climate Change Impacts on Livestock: Adaptation and Mitigation*, Springer, New Delhi Heidelberg New York Dordrecht London, pp. 285–301.

Garg, M. R., Sherasia, P. L., Bhanderi, B. M., Phondba, B. T., Shelke, S. K. and Makkar, H. P. S. (2013). Effects of feeding nutritionally balanced rations on animal productivity, feed conversion efficiency, feed nitrogen use efficiency, rumen microbial protein supply, parasitic load, immunity and enteric methane emissions of milking animals under field conditions. *Anim. Feed Sci. Technol.* 179: 24–35.

Garg, M. R., Sherasia, P. L., Phondba, B. T. and Hossain, S. A. (2014). Effect of feeding a balanced ration on milk production, microbial nitrogen supply and methane emissions in field animals. *Anim. Prod. Sci.*, 54: 1657–61.

Garg, M. R., Sherasia, P. L., Phondba, B. T. and Makkar, H. P. S. (2016). Greenhouse gas emission intensity based on lifetime milk production of dairy animals; as impacted by ration balancing programme. *Anim. Prod. Sci.* (in press).

Gerber, P. J., Steinfeld, H., Henderson, B., Mottet, A., Opio, C., Dijkman, J., Falcucci, A. and Empio, G. (2013). Tackling Climate Change Through Livestock – A global assessment of emissions and mitigation opportunities, Rome, Food and Agriculture Organization of the United Nations FAO.

Gerosa, S. and Skoet, J. (2012). Milk availability: trends in production and demand and medium term outlook. ESA Working Paper No 12-01, February 2012, Agricultural Development Economic Division, FAO: www.fao.org/economica/esa

Havlik, P., Valin, H., Herrero, M., Obersteiner, M., Schmid, E., Rufino, M. C., Mosnier, A., Thornton, P. K., Bottcher, H., Conant, R. T., Frank, S., Fritz, S., Fuss, S., Kraxner, F. and Notenbaert, A. (2014). Climate change mitigation through livestock system transitions. *PNAS*. www.pnas.org/cgi/doi/10.1073/pnas.1308044111

Herrero, M., Havlík, P., Valin, H., Notenbaert, A., Rufino, M. C., Thornton, P. K., Blümmel, M., Weiss, F., Grace, D., Obersteiner, M. (2013). Biomass use, production, feed efficiencies, and greenhouse gas emissions from global livestock systems. *PNAS* 110 (52): 20888–93.

IES (2010). ILCD Handbook. International Reference Life Cycle Data System. General guide for Life Cycle Assessment – Detailed guidance, Ispra, European Commission, Joint Research Centre. Institute for Environment and Sustainability (IES).

Ketelaars, J. J. M. H. and Tolkamp, B. J. (1992). Towards a new theory of feed intake regulations in uminants. 1. Causes of differences in voluntary feed intake: Critique of current views. *Livestock Prod. Sci.*, 30: 269–96.

Kristjanson, P. (2009). The role of livestock in poverty pathways. *Proceedings of Animal Nutrition Association World Conference*, 14–17 February 2009, New Delhi, India, pp. 37–40.

McDonald, P., Edwards, R. A. and Greenhalgh, J. F. D. (1988). *Animal Nutrition*. 4th Edition. Longman Scientific & technical, John Wiley & Sons, Inc, New York.

McMichael, A. J., Powles, J. W., Butler, C. D. And Uauy, R. (2007). Food, livestock production, energy, climate change and health. *Lancet* 370: 1253–63.

Mertens, D. R. (1988). Balancing carbohydrates in dairy rations. *Proc. Large Herd Dairy Management Con.*, Cornell University Anim. Sci. Mimeo 109, pp. 150–61.

National Institute for Animal Nutrition and Physiology (NIANP) (2012). *FeedBase*, Bangalore, 560–30.

Röös, E., Patel, M., Spangenberg, J., Carlsson, G. and Rydhmer, L. (2016). Limiting livestock production to pasture and by-products in a search for sustainable diets, *Food Policy* 58: 1–33.

Schader, C., Muller, A., El-Hage Scialabba, N., Hecht, J., Isensee, A., Erb K-H., Smith, P., Makkar, H. P. S., Klocke, P., Leiber, F., Schwegler, P., Stolze, M. and Niggli, U. (2015). Impact of feeding less food-competing feedstuffs to livestock on global food system sustainability. *J. R. Soc. Interface* 12: 20150891. http://dx.doi.org/10.10.1098/rsif.2015.0891

Shah, L. (2007). Delivering Nutrition. Power Point Presentation delivered at the CIGAR System Wide Livestock Program Meeting 17 September 2007 at ICRISAT, Patancheru.

Sharma, K., Pattanaik, A. K., Anandan, S. and Blümmel, M. (2010). Food-Feed Crop Research: A synthesis. *Anim. Nutri.Feed Technol.*, 10S: 1–10.

Singh, O. P., Sharma, A., Singh, R. and Shah, T. (2004). Virtual water trade in dairy economy. *Economic and Political Weekly* 39: 3492–7.

Steinfeld, H., Gerber, P., Wassenaar, T., Castel, V., Rosales, M., De Haan, C. (2006). *Livestock's Long Shadow: Environmental Issues And Options*. FAO, Rome, Italy.

Tricarico, J. M. (2016). Role of dairy cattle in converting feed to food. *Proceedings of the Tri-State Dairy Nutrition Conference Fr Wayne*, Indiana, USA, 18–20 April 2016.

Van Kernebeek, H. R. J., Oosting, S. J., Van Ittersum, M. K., Bikker, P. and De Boer, I. J. M. (2015). Saving land to feed a growing population: consequences for consumption of crop and livestock products, *Int. J. Life Cycle Assess.* 21(5): 677–87.

Vazquez, O. P. and Smith, T. R. (2000). Factors affecting pasture intake and total dry matter intake in grazing dairy cows. *J. Dairy Sci.* 83: 2301–9.

Village Dynamics in South Asia (2013). ICRISAT.org

Wirsenius, S., Azar, C. and Berndes, G. (2010). How much land is needed for global food production under scenarios of dietary changes and livestock productivity increases to 2030? *Agricultural Systems* 103: 621–38.

Chapter 3

Optimising the use of barley as an animal feed

David M. E. Poulsen, Queensland University of Technology, Australia

1 Introduction

Barley (*H. vulgare* sp. *vulgare*) is one of the six most globally significant feed grains. Wheat (*Triticum aestivum*), corn (*Zea mays*), rye (*Secale cereale*), oats (*Avena sativa*) and sorghum (*Sorghum bicolour*) round out the sextet. Triticale (× *Triticosecale*) is a relatively minor crop which adds to the pool of grain used as feed in some countries. Pulse and oilseed meals are often incorporated into livestock rations with the above-mentioned grains for balancing protein, fats and other nutritional factors.

More than 80% of the world's barley crop is used as feed (Blake et al., 2011) in both grain-fed and grain-supplemented livestock production systems. Barley is arguably the most adaptable of the feed grains. It is the most important feed grain in cooler climates where it can be more successfully grown than corn and the other feed grains (Hunt, 1996). Barley is used to feed ruminant animal species (beef cattle, dairy cattle, sheep and goats), monogastric animal species (pigs and horses), poultry and fish. Because of its outer husk, barley has a higher fibre and lower starch content than the other feed grains. However, barley starch is more degradable in the rumen than sorghum or corn starch (Hunt, 1996). While the percentage of the barley starch intake which is fermented in the rumen or hind-gut is similar to wheat and oats, it has a similar whole tract digestibility to corn starch and is intermediate between sorghum (low) and wheat and oats (high; Rowe et al., 1999). These, and other, properties of the barley grain make it a highly useful ingredient in feeding regimes as it can be strategically used to temper and moderate animal performance. Varying the structure and composition of barley grain, through exploiting genetic variation and

http://dx.doi.org/10.19103/AS.2019.0060.22

processing opportunities, means that a great deal of flexibility can be applied to the use of barley in feeding systems. This chapter will add to the growing body of work demonstrating that barley has been undervalued as a feed grain.

The second significant use for barley is for making beer by the malting and brewing industry. Malting barley selection is driven by highly defined and narrow grain specifications which are required to achieve optimal malting and brewing performance. To encourage production of grain which meets these specifications, malting barley is purchased at a market-driven premium price. While making little or no difference to the price premiums, malting barley and malt specifications can be further split into unofficial categories to meet the requirements of different brewing processes; namely, malt only, malt plus solid adjuncts and malt plus liquid adjuncts. It is a general rule that malting barley can only be produced using specific barley varieties which have been approved by industry for that purpose.

The remaining two mainstream uses for barley occur in the distillation industries (e.g. whisky, shochu) and for use as human food. Selling barley into the distilling industry works similarly to the malting/brewing sector, in that grain attracts premium prices to meet the demand under a strictly defined set of specifications. Food barley specifications and prices are set according to consumer and/or processor demands. Given that consumer markets for food barley span first-, second- and third-world economies, the specifications and prices for food grade barley are very wide ranging.

The premium prices for malting barley tend to drive the worldwide barley industry, despite the heavy use of the crop for feed. Consequently, most of the global crop is grown using accredited malting varieties. However, the strict requirements to achieve malting grade classification in most markets combined with very high Genotype × Environment (G × E) and Genotype × Environment × Management (G×E×M) effects on grain quality (Eagles et al., 1995; Molina-Cano et al., 1997; O'Brien, 1999; Fox et al., 2008) mean that a significant proportion of the annual crop fails to meet malting specifications. Grain which does not achieve the malting, distilling or food specifications, along with production from designated feed varieties, then tends to be amalgamated into the bulk commodity known as 'feed barley'. This has unfortunately created a generalised perception in most markets that 'feed barley' is essentially failed malting barley. However, the real situation is highly complex and feed barley is much, much more.

2 What is 'feed barley'?

The common use of the words 'feed barley' is a generic term used by industry to classify any form of the grain which is fed to livestock, poultry or fish in a managed production system. It is, in effect, a catch-all term. Different classification names are used for feed barley segregation around the world; these include 'feed barley' in Australia, 'general-purpose barley' in Canada and 'barley' in the United States. Most jurisdictions are also likely to have two or more sub-categories of 'feed barley' which are segregated on the basis of physical attributes which are believed to be related to the energy content and feed value of the grain.

Malting barley is classified as 'malting barley' worldwide. It is mostly stored and traded in its unprocessed form as single variety lots. Malting is a managed bulk grain germination process and is best undertaken in single variety batches to reduce the risk of processing complications from uneven germination and grain modification rates. Finished batches of malted barley, otherwise known as malt, can be blended to brewers' specifications with less risk of affecting the end-product quality.

The added costs of segregating into single variety or style loads mean that most of the globally traded feed barley consists of blended bulk grain. In general, loads of feed barley will usually consist of a combination of grain types and varieties. Multi-variety bulks may be traded on an as-is basis, or, blended further to meet specific market requirements (e.g. blending to achieve a mean grain protein and/or test weight specification). It should be noted that there are exceptions in countries with highly sophisticated infrastructure such as Canada, where barley can be segregated to the 'boxcar' level. Where such segregation is available, it allows for some trading of more specific feed barley batches with particular attributes; for instance, classified to specific varieties or grain types including hulless, waxy, low-phytate and ruminant diet specific. Another situation where more specific batches of barley are traded as feed is when direct grain farm to livestock producer contracts are in use. Even malting barley has been traded for feed use under such contracts (Kym McIntyre [Queensland Barley Industry Development Officer], pers. comm.).

For the most part, the selection and development of varieties released as feed barley has been less constrained by specific quality parameters and more focussed on factors influencing grain production and crop protection. Yield has been the primary drive feed variety selection, with the attention to feed quality mostly limited to choosing genotypes with plump and heavy grain. Grain size and density has been said to lead to better feed quality by indicating more starch per grain, and better malting quality through enabling a high extract when the enzymes which drive good modification are also present (Burger and Laberge, 1985). This has led to the overall situation where feed barley varieties worldwide tend to outyield malting varieties (Blake et al., 2011). However, there is no doubt that yields of both feed and malting varieties have continued to increase. Friedt (2011) indicates that overall barley productivity has risen by an annual rate of 1–2% per annum due to both genetic and production system gains, and, that this has included significant gains in malting quality as well as improvements in yield stability through enhanced agronomic traits and genetic resistances to diseases and pests.

The sources of feed barley varieties have been both releases from dedicated high-yielding feed breeding programmes and as spin-off releases of high-yielding but lower-quality material from dedicated malting barley breeding programmes. Because malting barley commands a premium over feed barley in most countries, farmers have a perceived incentive to continue to plant malting varieties. Variety choice for sowing becomes a calculated risk, with farmers needing to consider the risks of not achieving malt grade against the potential of achieving the maximum yield advantage of the competing feed varieties. Blake et al. (2011) provided an elegant summary showing how the malting premiums drive the profitability of malting over feed barley feed in the United States. The data indicated that feed barley varieties need to achieve a minimum of 10% in yield advantage over malting varieties to become financially competitive. Our own experience in eastern Australia confirmed this information (McIntyre, pers. comm.), and even suggested that the yield advantage alone required to encourage farmers to switch to growing feed types was at least 15–20%, this being more than the offered premium for malt. The net effect of this market-led perception is that malting barley varieties continue to dominate the worlds production. With increasing economic pressures and funding shortfalls, the number of feed barley programmes focussing on yield improvement alone appears to have significantly decreased.

3 What do we want from 'feed barley'?

The aim of livestock production systems is to optimise the balance of feeding efficiency with average daily output of produce from the species in question, regardless of whether the end-product is meat, eggs, wool, milk, breeding capacity or performance. The feeding regime is, therefore, critical as it can substantially impact feeding efficiency, animal production and the produce end quality. As well as considering the grain, there needs to be an understanding of the animal as this is a complex biological system with multiple layers of genetic, environmental and management effects in play. Animal species is obviously important, as is genetic variation for a multitude of traits within the species. The age of the stock will also have an impact on how a diet is digested and so specific feed formulations will usually be applied at the various growth and development stages for each species.

Intensive livestock production systems tend to use blended feed rations to provide diet consistency over long time periods, as it is understood that changes of diet in those systems can set back productivity while the stock adapt to the new feed source. Free range and mixed production systems will typically use whole or processed grain as a supplementary feedstock to make up for gaps in the natural foraging of the stock or to 'finish off' stock growth prior to sale.

Barley is used as a source of energy and protein when fed in livestock production, with energy being the most important of the two traits. It may be fed as whole grain, rolled or flaked, processed and blended into a mash type ration, or, other styles of product such as extruded pellets. To work out what is required from a feed barley, it is best to start by defining the characteristics of a malting barley as these are more precisely known.

Classification of a barley lot as either malting or feed grade first depends on its variety of origin. Malting grade barley can only come from an industry approved and certified malting variety. Modern malting barley varieties have been bred for the physical and biochemical attributes required for use in industrial malting and brewing industries. To be classified as malting, they have passed the stringent quality testing regimes of both the breeding programmes and the industry.

Malting barley varieties have largely been selected for their genetic capacity to efficiently provide sugars and starches to feed yeasts in the brewing, or distilling, processes. For most of the 10000 years or so that barley has been in cultivation and used to make beer, this has typically been understood as being lines with plump and uniform grain, high malt extract and quick and even germination when malted (Sparrow and Doolette, 1975). In recent history, however, the inherent attributes malting varieties must possess are far more defined and include high starch levels, appropriate levels of germination enzyme activity to modify starch and protein as required in the brewing process, moderate protein levels (e.g. 8–11%), large (but not too large) grain size, low-to-moderate grain dormancy and low levels of residual biochemicals such as -glucan and unattractive flavour compounds which interfere with the brewing process and products. Some variation in the ranges of these traits is required to meet the needs of different brewing styles; however, all modern malting varieties will have been selected for these attributes.

The grain produced from malting barley varieties can be rejected from the premium classification for many reasons. The variety identity must be declared; noting that the grain accumulator usually has access to physical or biochemical tests which can prove identity in cases where there is doubt of authenticity. The barley must then meet the physical malting classification specifications as demanded by the market and regulated within the country of production. Reasons for rejecting barley from malt classification include, but are not limited to, protein outside of the specified range for malting, small grain, low test weight, weather damage, grain colour, presence of black point, the presence of fungal contaminants, contamination with other barley varieties, weed seeds or soil. If rejected as malt at a grain receival depot, the barley will usually be added to the appropriate bulk feed grade storage according to its sub-classification. It is also noteworthy that some of the above-mentioned reasons for rejecting malt barley can also have negative impacts on the value of the grain for feed purposes.

As malting barley specifications have become more defined, there have also been efforts to more clearly define attributes contributing to better barley feeding quality. Bhatty et al. (1975) stated that data from a study with 16 barley cultivars showed that good malting quality in barley is not incompatible feed value, specifically with high digestibility of energy (D) and digestible energy (DE), as determined by mouse feeding. Other studies with livestock of various kinds continue to demonstrate that malting barley quality is largely consistent with feed quality (Hart et al., 2008; Fox et al., 2009), with some exceptions that will be discussed later.

Feed grade barley is classified by relatively few characteristics, when compared to malting barley. Bulk density measurements, test weight or volume weight, have been the main assessment standard in many markets. However, most evidence from studies indicates that bulk density does not correlate well with animal performance from varying sources of fed barley (Grimson et al., 1987; Hunt, 1996). The ability to undertake research into feed barley quality has been constrained by several factors, including the complexity of dealing with multiple target species, the cost and ethical considerations involved in live animal feed-quality assays and a significant lack of funding support. As a consequence, most so-called feed barleys released prior to the 1990s were subjected to any form of selection for feed quality apart from grain size or bulk density measurements. Since then, however, there has been a trend towards the release of some varieties of feed barley which have been shown to address specific animal feed requirements.

Substantial genetic variation has been reported in barley for a range of attributes which may have impact on the dietary value of the grain for different species. One of the key characteristics identified for feed barley breeding aimed at monogastric species has been the hulless trait, controlled by the recessive *nud* gene. In hulless barley, the hull is less firmly attached to the kernel than in the conventional genotypes and consequently becomes detached during threshing. This results in a barley which has a higher starch content and DE per unit of volume or weight. Reports of other notable feed-related attributes have included high amylose or low amylopectin (Merritt, 1967), high-lysine (Munck et al., 1970; Oram and Doll, 1981) and high-lipid content (Parsons and Price, 1974). Each of these have been investigated further as breeding targets. An extensive study of the USDA barley core collection demonstrated the existence of a substantial amount of exploitable genetic variation for barley feed quality (Bowman et al., 2001).

Apart from yield and specific quality attributes, attention must be given to the disease and pest resistance traits which are required for crop protection. These have been given equally high priority in selection of both feed and malting varieties. This is warranted by the combination of negative effects which diseases and pests may have on yield, production stability, the starch and other quality attributed of starch content of grain from affected plants. In some cases, resistances are critical to protect from other negative biotic effects of pathogens, such as the potential presence of toxins in grain affected by Fusarium Head Blight (FHB).

To summarise this section, when considering what is required from a feed grain, in this case barley, it is important to consider the entire system from sowing the seed through to the end-product from the livestock production. In the modern world where input resources such as water, land and fertiliser are increasingly limited and precious, this means that optimisation of the entire supply and value chain is the only logical way forward to achieving an outcome for sustainable production and consumption. What we are really looking for is the feed grain system which gives us the best and most efficient animal production per hectare of sown ground.

4 Optimising feed barley use

4.1 Optimising feed barley production

G × E × M interactions have a significant effect on the quantity and quality of barley produced in a crop, with yield, starch content, grain protein concentration and grain size distribution being highly vulnerable to heat and moisture stresses (Eagles et al., 1995). The impact of heat and moisture stress on starch accumulation in the grain is most severe in the early stages of grain filling, while protein is affected by stress applied from early through to late grain fill (Savin and Nicolas, 1999). Nitrogen application can also have significant impacts on these factors, including yield, protein and starch concentration (Perrott et al., 2018b). Genotype interactions with both the seasonal environment and management practices can be highly significant, and the relative maturity class of the variety can be a major co-factor in influencing the genotypic responses (Eagles et al., 1995). These G × E × M effects create a large amount of variation in the feed value of barley. As the variation in barley grain protein and energy content is demonstrably large, and arguably greater than seen in other feed grains, it becomes essential to analyse barley which is to be used for either feed or malting purposes (Hunt, 1996). A study of over 1600 feed barley samples from the northwest United States demonstrated highly variable chemical composition and ruminal digestibility, with source, location and variety accounting for approximately equal variability in two data sets (Reynolds et al., 1992). Therefore, to be able to produce feed barley, it becomes essential to optimise the relevant production system for every growing area. A simple 'one-size-fits-all recipe' does not work due to broad suite of environments in which barley is produced. Regional studies to determine optimal growing conditions need to be routinely conducted in order to correctly sample the relevant genotypes, soil and environmental conditions. All of these production inputs vary over time, with changing environmental conditions being foremost in many minds. Barley production is discussed elsewhere in this book, and for the purposes of this chapter a recent sample study (Perrott et al., 2018a,b) is referenced as an example of the ongoing agronomic research required to achieve optimal feed grain production.

4.2 Breeding feed barley varieties

For reasons described previously, continuing to pursue the 'yield alone' (incorporating disease resistance) strategy for the development of feed barley varieties is commercially questionable in the current world market. High-yielding feed barleys have generally only succeeded at commercial levels in areas where the production of malting barley was somehow constrained. For instance, the variety Galleon was highly successful in certain South Australian production zones for many years due to its resistance to Cereal Cyst Nematode (CCN) and the variety Grout was popular in Queensland for several years due to its ability to produce a decent crop under drought conditions which had significantly impacted the available malting varieties. In both case, as new malting varieties with equivalent agronomic traits and effective genetic resistances were released, the area sown to the feed varieties rapidly decreased and the malting varieties became dominant.

Nevertheless, the breeding for 'just yield' strategy has been proven as useful in improving the overall barley gene pool, as genetics from the high-yielding material has made its way back into malting barley development and helped to increase the yields of malting barley varieties. Researchers have investigated the strategy of using molecular marker techniques to introduce malting quality genes and quantitative trait loci (QTL) into feed barley backgrounds for the express purpose of rapid yield improvement of malting types (Vasos et al., 2004).

The availability of a substantial level of genetic variation in barley for improvement of the crops feed quality has already been mentioned. This resource has led to the development of breeding programmes in which specific feed-quality traits have been targeted as part of the overall variety improvement strategy. Breeding programmes of this type have been established at least in North America, Europe, Australia and Asia. The programmes tend to either focus on developing feed types for ruminant species or for monogastric species.

Barley hull content is known to be a major contributor to the lower digestibility of barley grain when compared to wheat, oats and maize. In the monogastric industries, this is being addressed through the increasing use of hulless barley varieties. For hulled barley, the hull content varies from 7% to 25% among the full gene pool including two-row and six-row barley genotypes, whereas two-row barley as a subclass commonly displays lower hull content with a mean of 10% (Evers et al., 1999). Genetic variation within the modern cultivated two-row barley gene pool certainly exists; however, it is relatively narrow as demonstrated by the hull content range of 9–11% observed in six North American two-row varieties (Du et al., 2009).

The pork industry seems to have been the most proactive in supporting the development of specific feed varieties. This is reflected in results as, arguably, the most successful quality-specific feed barley varieties have been those produced for the North American pork industry by the breeding programmes operated by Alberta Agriculture and the University of Saskatchewan. These have included hulless, waxy, waxy-hulless and low-phytate varieties. The success of these barleys has been driven in part by both the investment of the pig producers in the breeding programme and the preparedness of the pig producers to pay a premium for grain that they want (Jim Helm, former Alberta Agriculture barley breeder, pers. comm.).

Another well-publicised and moderately successful specific purpose feed barley variety has been Valier (Blake et al., 2002), released by the Montana State University specifically to improve the performance of feedlot cattle. Valier was selected for the combination of excellent agronomic characteristics with improved cattle feeding performance, the latter being demonstrated in calf feeding trials comparing the variety as a breeding line against both of its parents (Boss et al., 1999). A comparative analysis of the nutritional value for cattle of Valier versus the predominant malting variety Harrington was conducted (Yu et al., 2003). The authors concluded that the nutritional advantage of Valier was highly associated with the characteristics of the feed after the barley was dry rolled. Despite chemical and nutritional composition of grain from each variety, there was little difference in rumen degradation characteristics between the two barleys.

The potential for developing even better barley varieties with improved generic and/ or specific feed barley qualities is very high. While there is still a lot of research and development work to do, increased knowledge about animal dietary requirements and combined new breeding and selection technologies such as near-infrared spectroscopy (NIR) and molecular markers, have given us the capability to be highly focussed on making the necessary improvements to optimise feed barley performance.

4.3 The Premium Grains for Livestock Program: a model for developing the knowledge required to optimise the use of grains in feeding livestock

Arguably, knowledge is the most powerful weapon in a plant breeders arsenal. It enables the breeding team to make decisions which significantly increase the probabilities of finding the best genetic combinations for each situation. The breeding programmes must, of course, be market driven and working closely with the industry which represents their target market. Industries must be similarly driven and work together to achieve the best possible mutual outcomes, instead of competing against each other to squeeze the last dollar out of the value chain for themselves. This approach has worked for the Canadian pork industry, as discussed previously. The Australian grains and livestock industries have also been working together to gain a better understanding of the requirements for optimal feed grain performance in animal production. The Premium Grains for Livestock Program (PGLP) was a milestone project conducted in Australia, with the aim of characterising all available feed grain sources along with livestock feeding, digestion and growth factors to generate a major source of knowledge for use by the industry in optimising the national grain-fed animal value chain (Black, 2018). Barley, as Australia's third most commonly produced feed grain, featured throughout the PGLP studies and publications.

For ease of access, a series of papers from the PGLP Stage I were collated and published in a special edition of the *Australian Journal of Agricultural Research* (AJAR, 1999, Vol 50, Issue 5). This volume is a seminal source of information for researchers and others seeking information on the topic of optimising the utilisation of feed grains in the animal industries. The journal contains a number of useful articles, including important reviews of the various *in-vitro* and *in-vivo* feed-quality testing protocols in use for ruminants (Hogan and Flinn, 1999; Kitessa et al., 1999; White and Ashes, 1999), pigs (Moughan, 1999), poultry (Farrell, 1999; Ravindran and Brydan, 1999) and general (Peterson et al., 1999; Wrigley, 1999) assessment activities.

Kaiser (1999) presents arguments in favour of reducing grain processing costs and improving animal performance by increasing the digestibility of whole grains in cattle through the dual strategies: (1) breeding grains are more readily damaged during mastication and with more digestible seed coats, and (2) treatment of the grains with fibrolytic enzymes or chemicals such as NaOH.

A review of the effects of the common practice of combining grain with forage in ruminant feeds to improve animal performance (Dixon and Stockdale, 1999) concluded that both positive and negative effects were possible. The positive effects were mostly related to the grain providing additional nutrients which were deficient in the forage, or by stimulating forage digestion when added at low percentages of net intake. Negative effects were associated with lower intake and digestion of forage components, but could be minimised by careful selection of compatible grain and forage combinations. Barley, along with wheat and oats, is more likely to contribute to negative intake and digestion than sorghum or maize because of the risk of lower rumen pH from the highly fermentable starch. On the other hand, the authors suggest that a satisfactory compromise forage mixture using grain processed only to the extent necessary to be adequately digested in the entire gastrointestinal tract (GIT) will usually be easier to achieve with rapidly fermented grains such as barley than with slowly fermented grains such as sorghum.

In grain-fed production systems, the types of grain and processing need to be balanced to achieve optimal production and minimise risks of low pH issues in the rumen and/or intestines. Manipulating feed rations by adjusting mixes to take account for the relative digestibility of wheat, oats, barley, maize and sorghum and how this can be managed by the amount of rolling and steam flaking can be used to manipulate the main sites of starch digestion within the GIT of various species (Rowe et al., 1999).

The PFG Program highlighted that all feeding and breeding strategies which rely on grain quality as a key piece of knowledge must have rapid, cost-effective and accurate analytical tools in order to succeed (Wrigley, 1999). Past means of assessing have included visual inspection, physical characterisation and detailed laboratory chemical analysis. Visual and physical assessment tools provide an indirect, rudimentary and largely inaccurate means of estimating feed quality. On the other hand, laboratory analyses provide detailed information but can be expensive and take a long time to produce results for comparison of multiple samples. One of the most game-changing technologies for rapid and accurate grain-quality assessment is NIR, an imaging system which is now being used in grain receival depots, feed manufacturers, plant breeding programmes and research laboratories. One of the outputs from Stages I and II of the PFG was the release of a suite of NIR calibrations developed to give feed-quality predictions across the full range of grains being utilised in Australia (Black, 2018). Stage I of the PGLP produced a set of calibrations useful for characterising the DE of cereals to be used for pig feed (van Barneveld et al., 1999). By the end of the second stage of the Program, calibrations had been released to industry for use with whole and milled grains, on a dry matter or as fed basis and for cattle, sheep, pigs, broiler chickens and layer chickens. These calibrations have been useful to grain accumulators, feed manufacturers and livestock producers. However, they are of relatively little use in barley genetic improvement since, due to the nature of NIR technology and the physical and chemical variation between the grain of crop species, only the use of barley specific algorithms is likely to achieve significant feed-quality gains through plant breeding (Glen Fox, pers. comm.).

4.4 Near-infrared spectroscopy (NIR) and molecular markers: modern selection tools for improving feed quality

Fast, cost-effective and reliable screening techniques which allow the assessment of grain from large numbers of breeding lines is a critical need for any breeding programme targeting quality improvement for any purpose. The non-invasive predictive NIR technology is now widely adopted by crop breeding programmes as a means of rapidly screening for a diverse range of attributes, from predicting relatively simple traits such as protein and starch content, through to the significantly more complex traits such as DE and ADG in

60

Optimising the use of barley as an animal feed

different livestock species which would usually rely on both *in-vitro* and *in-vivo* trials to confirm (Wrigley, 1999; van Barneveld et al., 1999; Fox et al., 2011).

One of the earliest uses of NIR to make feed-quality selections in barley occurred in the Alberta Agriculture breeding programme, with pigs as the primary target species (Jim Helm, Alberta Agriculture barley breeder, pers. comm.). The Queensland barley programme in Australia was also a relatively early adopter of NIR to select for feed quality, developing calibrations and using the technology to predict traits which had been identified from studies at Montana State University (Surber et al., 2000) as pertinent to developing better-quality barley varieties for feedlot cattle. By 2006, the Queensland barley breeding programme was testing grain samples harvested from over 10 000 plots within a 10-week period (Poulsen, 2006). This capability was expanded further as the breeding evolved over the next 5 years (Franckowiak, 2011). Like the physical and chemical traits which it predicts, significant variability can be observed in barley NIR trait measurements between genotypes, environments, crop management systems and the various combinations of interactions, particularly G × E (Fox et al., 2011). This reinforces the need for breeding programmes to constantly update and refine calibration sets to account for the gene pool/s in use and the test environments. As well as its use for selection per se, studies have shown that NIR can be used as a reliable means of developing data sets for the identification of molecular markers linked to single genes and QTL conferring barley malting and feed-quality attributes (Gous et al., 2012b).

Plant breeders also have access to molecular marker technologies to assist with both speeding up and improving selection efficiency in crop breeding programmes. The first two molecular marker maps of barley were developed using European and wild barley (*Hordeum spontaneum*) germplasm (Heun et al., 1991; Graner et al., 1991). These were a precursor of what was to come, and major steps with the development of barley molecular marker technologies occurred with the two major multi-organisational initiatives of the North American Barley Genome Mapping Project (NABGMP) and the Australian Winter Cereals Molecular Marker Program (AWCMMP), the latter which featured genome mapping for both barley and wheat. The significance of these initiatives, in which the barley research between the two programmes was also linked, was that they used the combined resources of breeding programmes, research teams and laboratories from multiple organisations, growing regions and countries to produce an unsurpassed series of plant genetic maps using multiple marker types and linking these to genes, the expression of traits and QTL through the use of a globally collected phenotypic dataset.

The first NABGMP high-density map from the cross between the six-row feed variety Steptoe and 6-row malting variety Morex (Kleinhofs et al., 1993) was one of the most significant steps in modern barley improvement, in that it provided a framework to study, identify and map genetic variation for a broad set of traits. Grain and malting quality traits were quickly added to the Steptoe × Morex map (Hayes et al., 1993), and subsequently to the second major NABGMP population from the two-row malting barley cross Harrington × TR306 (Mather et al., 1997; Igartua et al., 2000).

The Australian-led barley mapping project was somewhat different from the NABMMP, in that multiple barley populations were mapped to represent a broad selection of the gene pool being used by the several state-based breeding programmes which were in operation during the period that the AWCMMP was initiated (Karakousis et al., 2003a,b; Barr et al., 2003a,b; Read et al., 2003; Cakir et al., 2003; Pallotta et al., 2003; Long et al., 2003). The individual populations were mapped and then used to construct the first published consensus map of the barley genome (Karakousis et al., 2003c) based on five of

© Burleigh Dodds Science Publishing Limited, 2020. All rights reserved.

the AWCMMP barley maps and using Singe Sequence Repeat (SSR), Random Fragmented Polymorphic DNA (RFLP) and Amplified Fragmented Polymorphic Marker (AFLP) marker types.

The AWCMMP mapping populations included feed and malting barley parents to ensure that phenotypic data for a broad range of grain, malting and feed-quality attributes could be captured and mapped. The 'feed' parents were not speciality feed varieties, but were bred either for high yield in specific target environments or had been released as potential malting varieties but later classed as feed after being rejected by failing to meet malting and brewing industry testing protocols. Therefore, most of the grain-quality traits mapped through the AWCMMP barley populations were either the general quality traits required to meet receival specifications (grain size, protein content) or more specific malting attributes.

A comprehensive study conducted as part of the AWCMMP and using a combination of Australian and International barley maps indicated that the detectable QTL for grain weight and size were, at that time, all associated with barley developmental genes (Coventry et al., 2003). These genes included the Ppd-H1 locus (long day photoperiod response) in European mapping populations and studies, the eps2 locus (developmental rate determination) in Australian mapping populations and studies, the sg1, sg2 and sg3 vernalisation loci and the allelic denso and sdw1 semi-dwarfing loci found in two- and six-row barleys, respectively. Because of the high level of influence of those genes on grain size, and the significant G × E effects which affect their expression, the mapping of non-developmental genes affecting grain size requires a specific mapping population where the major developmental loci are not segregating. However, the use of molecular markers to select for grain size may well be unnecessary, as it has been demonstrated that grain size is highly heritable (Fox et al., 2006) and that the combination of modern statistical analysis with traditional breeding and selection methods can be highly effective in selecting large-grained genotypes.

Genes contributing to the variation in grain protein have also been mapped (See et al., 2002; Emebiri et al., 2003). While both of the studies were intended to develop marker systems for selecting low protein varieties to increase the chances of growing malting grade barley in less favourable conditions, they could also be used to enhance feed barley breeding strategies.

The hulless barley gene nud has been mapped to the long arm of chromosome 7H (Fedak et al., 1972), and linked to microsatellite markers (Becker and Heun, 1995). Nevertheless, it took some time for additional specific feed-quality traits to be investigated by either the NABMMP or AWCMMP. A population derived from a cross between the two-row hulled feed barley Valier and six-row hulless Swiss landrace PI370970 has been used to map and identify multiple QTL for in-situ dry-matter digestibility (ISDMD) and post dry rolling particle size (PS), with 35–62% of the variation in ISDMD and 26–53% of the variation in each trait, respectively, linked to the vrs1 locus which controls the two-row/six-row trait (Abdel-Haleem et al., 2010b). Thirty main effect and four epistatic QTL were also mapped in the Valier × PI370970 population as related to the expression of ADF, starch content and protein content (Abdel-Haleem et al., 2010a). Again, there were strong associations with the feed-quality traits and a major gene. In this second study of the population, major QTL for ADF and starch content were associated with the nud gene which confers the hulless trait. This association between major developmental traits and QTLs reflected information from previous studies of the impact of the genes on physical and chemical grain characteristics.

Thirty-two main effect and five epistatic QTL for cattle feed-quality characteristics were added to the Steptoe × Morex mapping data; namely, QTL for crude protein, ADF, starch content, PS, ISDMD for cracked and ground grain and the starch digestibility of ground grain (Abdel-Haleem et al., 2012). An Australian study, using the Tallon × Scarlett AWCMMP mapping population and fistulated cattle, identified one QTL associated with digested protein and two QTL associated with DMD (Gous et al., 2012a). The genetic associations as described previously indicate that selection for improvement of specific feed barley quality attributes can be effectively managed through a combination of traditional section for major developmental genes and marker-assisted selection for more subtle genetic effects. However, an advantage of marker-assisted selection is that it can be applied in seedlings as a tool to select or eliminate heterozygotes with otherwise undetectable recessive genes and homozygotes before the plants need to mature to the age where they are expressed.

4.5 Processing

Whole grains can be readily fed to sheep and chickens; however, as whole grain can be difficult for most other animals to digest and the barley hull has been shown to reduce the grains digestibility when compared to wheat, some form of grain processing can be used to improve the accessibility of barley starch and protein within the GIT (Rowe et al., 1999). The best type of processing to use is generally determined by the animal species, and particularly, adjusting the feed mix so that the main sites of grain digestion within each species GIT are optimised to make the most of the nutritional value of the grain.

One of the simplest processing methods to improve digestibility is dry rolling of the whole grain, which can then be fed as is or blended into a ration mixture with forage material and/or seed meals. Dry rolling has been shown to significantly increase grain digestibility, with the effect of dry rolling in barley being significantly greater than seen in most other grains. For example, an Australian experiment using 2-year-old steers demonstrated an improvement in digestibility after dry rolling in barley, wheat and oats of 41.9%, 27.7% and 4.0%, respectively (Toland, 1976). An alternative to dry rolling is hammer milling, which similarly uses physical force to break the grain, thereby exposing the starch in the endosperm by both reducing the effect of husk and increasing the access to starch by the larger surface area-to-volume ratio of the rolled or milled grain particles. Milling has a significantly more disruptive effect on the grain, in particular resulting in small PS. Care must be taken not to over process grain, because rapid digestion as a result of excessive small particles or flour in the feed is likely to lead to low rumen pH and subsequent acidosis and low-fibre digestibility in ruminant species (Rowe et al., 1999). It is important, therefore, to adjust the rolling or milling equipment to achieve optimal PS to match the characteristics of each separate load of grain.

Steam flaking can also significantly improve barley digestibility. Most studies show that steam flaking offers little advantage over dry rolling in barley (Rowe et al., 1999). There appears to be little difference in digestibility in cattle between rolled and steam flaked barley, apart from the first month of feeding where steam flaking may offer a small advantage (Grimson et al., 1987). The downside of such treatments to markedly increase the digestibility of starch is that they can contribute to a rapid breakdown of starch in the rumen and the consequent problem of acidosis, leading to poor health and performance in cattle (Blake et al., 2011). Sheep are also vulnerable to acidosis, which will affect both weight gain and wool production (Hynd and Allden, 1985).

In general, the optimal way to process barley for ruminants is dry rolling (Rowe et al., 1999). Steam flaking will improve overall digestibility; however, it comes with a higher risk of high fermentation and acidosis. Additionally, the economics of the two systems means that dry rolling is likely to be more cost effective due to the additional processing costs of steam flaking for relatively little, if any, gain. The opposite holds for pigs, where steam flaking results in higher intestinal and overall digestibility of grains, and gut fermentation is not as significant a risk to the animal. As indicated previously, poultry can be fed whole grain; however, steam flaking will improve the overall digestibility when compared to either whole or cracked grain feed preparation methods.

A minor use of barley in stock feeding is the addition of small amounts of whole or rolled barley to silage as a means of increasing the digestibility of the forage component. This practice has been shown to significantly improve cattle feeding performance when compared to a silage-only diet (Jacobs et al., 1995), and is particularly useful in dairy cattle production. A 2011 study showed that barley, corn-dried distillers grain, wheat-dried distillers grain and wheat middlings when fed with corn silage were interchangeable in such a feeding system without adverse impacts on production (Holtshausen et al., 2011).

Another form of barley modification for feed rations, and one which is not available to most other grains, is the use of spent brewers grain. The malting and brewing process significantly modifies the barley, with the resulting material having a lower carbohydrate content and elevated protein levels. Because the spent grain is usually wet, it is not suitable for long-distance transport and this suggests that such material is more likely to be useful to industries closer to the urban centres where the breweries are located. Studies have shown that spent brewers grain is a very useful feed source in dairy production, and can be used to replace rolled barley in diets incorporating a mixture of hay and grain. Daily yields of milk, protein and non-fat milk solids have been shown as significantly higher from diets incorporating brewers grain over rolled barley (Valentine and Wicks, 1982).

5 Understanding and optimising feed barley quality for different livestock species

As has already been discussed in this chapter, barley is fed to diverse species. Cereal grains are primarily fed to animals as an energy source (Rowe et al., 1999), with the protein content and type being an important secondary condition. Hulled barley is also recognised as a source of dietary fibre, particularly for ruminants. The different dietary needs of each species, together with the varying dietary needs of animals and birds in different age categories, means that it becomes very complicated to define a specific set of criteria for the ideal and generic 'feed barley' (Bleidere and Gaile, 2012). For example, hulled barley is arguably better for ruminant species as it is easier to regulate gut fermentation due to the increased fibre, while hulless barley has become preferred for feeding monogastric species for a range of reasons including higher energy value and lower faecal output. Many grain-quality attributes have been identified, including physical, biochemical and nutritional characteristics. It thus becomes necessary that to optimise feed barley use, we need to understand what each type of animal requires from the grain.

5.1 Ruminants: cattle

Barley is the most significant feed grain used for cattle production systems in North America, and competes region by region with wheat and sorghum because of seasonal availability and crop distributions in Australia. The barley quality requirements for beef and dairy cattle are essentially the same, noting that there are significant differences between the feeding systems used for beef production, milk production and breeding. Most breeder cattle are fed on pasture and forage diets, with grain used for supplementation when the natural feed supplies are running short.

Beef cattle are either pasture or grain fed, with grain feeding being used mostly to 'finish off' animals which have been pasture fed through to a desired age before being introduced to a high-grain diet. This system has been shown to result in faster weight gain and higher quality meat. For example, 12-month-old Friesian steers showed a 31% faster growing rate on a predominantly barley diet than similar cattle fed predominantly on pasture, reaching a target fasted liveweight of 457 kg on average 69 days faster (Purchas and Lloyd Davies, 1974). Testing of meat from the slaughtered animals showed that the grain-fed steers also had significant higher fat content and greater tenderness measurements in specified cuts. A laboratory taste panel also indicated a more acceptable flavour from topside roasts of the barley-fed animals.

Numerous studies have been undertaken on the preferred barley attributes to optimise the performance of beef cattle. The optimal use of barley for ruminant feed can be related to both physical and chemical aspects of the grain. As previously discussed, large and plump grains are preferred because they contain a high percentage of starch compared to husk. Because of the presence of the husk, hulled barley contains 10–15% less starch per unit weight than maize, sorghum, wheat or oats (van Barneveld, 1999b). However, whole-tract digestibility of barley is superior to sorghum, similar to maize and slightly below wheat and oats. Anecdotal evidence suggests that the optimal barley grain protein levels for feeding cattle are believed to be in the range of 12–13%, as compared to the 8–12% protein range preferred for the malting process. The reasoning behind this being thought of as an optimal range for cattle is that the relatively indigestible hordein storage proteins are significantly higher in the high-protein barley, and as protein drops below 12% the nutrition available to support the rumen bacteria begins to decline. The endosperm protein matrix is also important in starch digestion, as it supports the starch granules, and as it degrades, it enables the ruminal bacteria to access the surface area of the starch. This is a primary factor in the lower digestibility of sorghum and corn as the protein matrices in those grains are less digestible than that found in barley, wheat and oats (van Barneveld, 1999b).

A high level of starch fermentation in the rumen is undesirable, as it can lead to a low pH and the consequent problem of acidosis. Acidosis can have significant negative effects on animal growth, milk production and overall animal health. In extreme cases, animal deaths will occur. Feedlots have learned how to manage acidosis, by using appropriate feed blends incorporating grain, silage or dry forage and pulse and oilseed meals. Systems vary between production areas and feedlot dieticians and managers learn how to manage acidosis risk accordingly, depending on the availability of produce at a cost-effective price. Anecdotal evidence suggests that this appears to correlate with the preference to use feed wheat as compared to feed barley, respectively, in Australian versus North American feedlots. A Canadian study investigating mixture proportions of silage and feed barley determined that optimum feed utilisation was obtained using more extensively processed barley combined with 3–6% of the dietary DM as barley silage (Koenig and Beauchemin, 2011). Fine-tuning of the diet formulations to decrease the risk of acidosis were shown as linked to decreased feed-conversion efficiency, and it was concluded that a decrease in the cost of animal mortality or sickness was necessary to justify the increased cost and use of low-risk acidosis.

Studies have demonstrated that there is a very high level of genetic variability for rumen digestibility between feed barley cultivars and as much as 70% of that can be attributed to variation in bulk density, starch content and kernel weight (Khorasani et al., 2000). Surber et al. (2000) identified a number of more specific feed-quality traits as important to beef cattle production, namely, low ADF, low ruminal DMD and large PS after dry rolling. In contrast with wheat, grain hardness in barley as conferred by the *hin b1* and *hin b2* genes appears not to be associated with feed quality, including PS (Fox et al., 2007c). Genetic variation in the above-mentioned traits is available (Bowman et al., 2001), and some breeding programmes adopted selection for these traits as part of their breeding strategy (Blake et al., 2002; Poulsen, 2006). The Canadian varieties CDC Dolly and CDC Helgason were identified as being good candidates as feed for cattle over four other Canadian varieties, on the basis of lower hull content, FA, PCA, fibres and moderate mean/median PS after dry rolling (Du et al., 2009).

5.2 Ruminants: sheep

As ruminants, sheep face similar issues as cattle when fed grains with readily accessible starch. One difference, however, is that sheep can be fed whole grain readily (REF). There appears to be little to no advantage in feeding processed barley to lambs (Sormunen-Cristian, 2013). Sheep are more likely to be pasture fed with supplemented grain, and are not typically housed in feedlot-like situations. Like cattle, sheep are vulnerable to low rumen pH and acidosis as a consequence of rapid digestion of starch. Acidosis can affect both animal growth and wool production. One way of managing the transition to barley-fed diets so as to minimise the risk of acidosis include a gradual introduction of the barley into the diet, administration of a course of antibiotics such as virginiamycin and the transfer of rumen fluid from well-adapted animals (Godfrey et al., 1992). An arguably better way to manage acidosis in sheep is simply to use a mixed-grain diet. The addition of slowly fermenting grains such as sorghum or corn into a barley diet has been shown to improve rumen pH, and by association, animal performance (Yahaghi et al., 2012).

Lambs appear to adapt faster to diets supplemented with barley as opposed to oats (Kenney, 1986; Sormunen-Cristian, 2013). Kenney (1986) found that liveweight was greater, but dressing percentage was lower on oat diets as compared to barley or wheat diets. In comparison, growth rate, carcass weight, slaughter percentage, kidney fat and carcass fat were shown to be higher with barley than oat dietary supplementation in the study conducted by Sormunen-Cristian (2013).

5.3 Pigs

Wheat and barley are commonly used in pig diets, with barley known to have lower average DE than wheat. In general, early studies demonstrated that pigs fed barley as compared to wheat tended to produce a similar to slightly higher liveweight gain despite a poorer efficiency of feed conversion (Lawrence, 1970). Barley, and other cereals, are most effective in pig diets when they have been processed, as all forms of processing significantly increase DE over that of grains (van Barneveld, 1999a).

A Canadian study assessing the dietary performance in recently weaned pigs of several wheat and one barley variety found that the barley-based diet had both lower DE and metabolizable energy; however, the difference was not reflected in live pig performance in an ad libitum feeding regime (Bowland, 1974). It was theorised that the higher lysine and threonine levels in the barley when compared to the wheat may have contributed to the good growth performance of the pigs on the barley diet, despite the lower DE than the wheat diets. More recently, a large feeding study using three feed mixes of soybean meal combined with either wheat, 'low-quality' hulled barley and 'high-quality' hulled barley echoed the aforementioned 1974 study and concluded that barley can full replace wheat in pig diets and achieve equivalent or better growth performance, despite the lower nutrient digestibility and energy value of the barley (Nasir et al., 2015). The addition of enzymes to pig diets can help to improve the digestibility of lower-quality barley (Clarke et al., 2018). By adding a combined -glucanase and -xylanase supplement to both low- and high-quality barley feed mixes, ADG, ADFI and nutrient digestibility were improved in the low-quality barley mix.

Data from analysis of over 125 barley cultivars has produced a range of DE estimates from 11.7 to 16.0 MJ/kg DM (van Barneveld, 1999a), this being an economically significant range when it comes to pig producers being able to optimise their productivity. The data suggested that energy digestibility of barley does not reflect gross energy (GE) content. The analysis of this, and data from other grains, indicated that cultural conditions and agronomic practices such as fertiliser application appear to have a greater influence on the availability of amino acids and energy than do the growing region or year.

Bhatty et al. (1975) recommended the desirability of developing hulless feed barley cultivars, especially for pigs and poultry, from mouse feeding trial results showing that the proportion of hull content being the major factor influencing the D and DE of barley. In their study, the hulless genotypes showed an approximately 8% increase in DE over the hulled types. Given that the average hull content of hulled barley is approximately 10% (Evers et al., 1999), this strongly suggests a direct effect from removing the hull and not as a by-product of the biochemistry of husk adhesion. Bhatty et al. (1975) also suggested that lipid content of barley may have a major influence on DE, while kernel size and lysine content had no effect on DE and were therefore viable breeding targets for improving DE, grain appearance and amino balance, respectively. Their follow-up study on pigs confirmed the above-mentioned recommendations with the DE of hulless barley being equal to or superior than corn in the new feeding trials (Bhatty et al., 1979). Additionally, the trials indicated that hulless barley could be a better feed for pigs than corn, because of the higher and superior protein content of the barley. The dietary advantage of hulless over hulled barley in the trial was shown to be 11.1% for D and 14.7% for DE. In another study, Mitchall et al. (1976) confirmed the advantages of hulless versus hulled barley by providing increased DE and total dietary protein content, even though there appeared to be a decrease in the protein digestibility of the hulless barley tested relative to its hulled isogenic type. By the 1990s, several hulless barley varieties had been released in North America and were in regular use as feed by the pork industry (Aherne, 1990; Darroch et al., 1996). Much of this work had been made possible by funding contributions made by the pig producers to breeding and research programmes, such as those run by Alberta Agriculture (Jim Helm, pers. comm.).

As well as being a source of energy for pigs, barley also contributes a substantial proportion of their protein requirements. Thacker et al. (1988) concluded that by combining an improvement in feed efficiency of hulless over hulled barley with the lower levels of soya-bean meal required with barley diets in order to meet the pigs' requirements for essential amino acids, sufficient incentive may be provided to encourage the incorporation of hulless barley in pig rations. However, the deficiency of some essential amino acids in barley protein, particularly lysine and threonine, necessitate the addition of protein supplements such as fish, meat or oilseed meal with or without pure methionine and lysine (Oram and Doll, 1981). Consequently, there have been plant breeding efforts to change the balance of amino acids in barley to render it more suitable for the porcine diet. Multiple high-lysine mutants have been discovered in barley; however, all are known to be associated with yield depression and lower-energy content due to reduced endosperm production. Genetic studies conducted in Australia and Europe, as reported by Oram and Doll (1981), identified potential biological pathways which could be exploited in breeding programmes to increase the yields of high-lysine variants and bring them closer to commercially viable yields. Exploitation of the material has been successful, with high-lysine varieties released and additional breeding lines showing greater commercial potential (Gabert et al., 1995; Darroch et al., 1996). This material has been shown to have a greater biological value for monogastric diets, as well as substantially reducing N excretion into the environment due to a reduction in the prolamin protein fraction (Gabert et al., 1996; Darroch et al., 1996). Transgenic approaches have been suggested to further increase the levels of lysine in feed and food barley (Shewry et al., 1994). The added attraction of the use of hulless varieties with higher levels of lysine, such as the variety Condor released in 1988, is that they may provide an economic benefit to pig producers because of their higher DE energy content when compared to hulled barley and wheat, reducing the need for costly fat supplements (Darroch et al., 1996).

Due to the cost, regulation and large amounts of grain required for conducting animal feeding trials, predictive tools had to be developed so that a greater degree of sophistication than just selecting for a hulless barley could be employed in pig-focussed breeding programmes. NIR is now the predictive tool of choice; however, earlier studies identified that apparent energy digestibility (AED) and total protein digestibility (TPD) assessment data from rats could effectively predict barley performance in pigs. *In-vitro* organic matter digestibility and mobile nylon bag dry-matter digestibility could also be used for AED. Gabert et al. (1996) used rats and young pigs to prove that high-lysine barleys would be beneficial to both monogastric animal health and performance and help to reduce N excretion into the environment.

The excretion of phosphorous into the environment from piggeries also has the potential to be a significant environmental problem (Pierce, 2000). This occurs because about 70% of the phosphorous found in barley is bound as phytate (Lott et al., 2000) and the necessary digestive enzyme phytase does not occur in pigs (Pointillart et al., 1984). There are two primary solutions to this problem, these being either addition of phytase to pig diets or development of barley varieties with lower phytate content. After the identification of low-phytate barley mutants (Larson et al., 1998), studies have concluded that the addition of this trait into feed barley varieties could offer significant advantages to the pig industry through lowering ration costs by reducing the need to purchase the inorganic phosphorous currently needed to provide a balanced diet (Thacker et al., 2003).

Shewry (2007) argued in favour of the use of genetic engineering to develop and release a competitively yielding low-phytate barley, given the yield disadvantages demonstrated by material developed from the two barley lines of this type, namely, Hiproly and RisØ (1508). A low-phytate variety named Piggy and released by Carlsberg in 1987 failed to achieve commercial success, possibly because the increased feed value did not compensate sufficiently for a small-yield penalty. Another conventionally bred low-phytate hulless variety, CDC Lophy-I, was released by the University of Saskatchewan's Crop Development Centre in 2008 (Rossnagel et al., 2008). Feeding trials conducted with the variety showed that pig phosphorous utilisation increased and that the grain had a similar DE by comparison to regular hulless barley (Ige et al., 2010). Lophy-I appears to produce reasonably competitive yields when compared with other hulless varieties grown in Canada (Gray et al., 2012). A second Canadian feeding study demonstrated higher ADG and nutrient digestibility than conventional hulless barley, and that addition of dietary phytase to the low-phytate barley diet improved feed conversion and nutrient digestibility (Woyengo et al., 2012).

5.4 Poultry

Although barley is a potential major source in energy in poultry feed, it has tended to be passed by in favour of other grains such as wheat, corn and sorghum. One of the reasons for this is that barley and wheat are also a source of anti-nutritive components which can have an impact on how effectively the complete diet is used by poultry (Hughes and Choct, 1999). Barley has been linked to a growth-depressing effect in broiler chicks (Anderson et al., 1960; Classen et al., 1985). Following multiple observations of the positive impact of using enzyme mixtures with barley-based diets in chickens, it has been determined that the -glucan polymer which occurs at high levels in barley and is associated with viscous intestinal contents that reduce gut digestion is largely responsible for the effect (Burnett, 1966; Mannion, 1981; Classen et al., 1988a). Additionally, this leads to sticky droppings which can lead to significant waste management problems in egg and poultry production systems.

Australian tests of commercial enzyme admixture to poultry feed in Australia (Mannion, 1981) confirmed that the inclusion of -glucanases and -amylases increased chicken feed consumption, digestion and weight gain and appeared to eliminate the occurrence of sticky droppings when compared to the control birds. Classen et al. (1988a) confirmed that the addition of -glucanases with barley grain rations increased growth and feed conversion in young birds and reduced the variability in performance created from feeding the chicks a range of barley samples with varying grain quality. The benefits from inclusion enzymes, such as -glucanase, have continued to be demonstrated in other studies (Campbell et al., 1993), leading to the regular inclusion of exogenous glycanases in broiler feeds for many years (Bedford and Morgan, 1996).

There is some evidence which suggests that the effects of barley -glucan/viscosity levels on growth rate may be age related. In Classens' et al. (1988a) experiments, the growth of chicks was slower in the first 21 days of feeding on the barley rations as compared to those on a wheat/corn diet; however, both diets then produced similar growth rates from day 21 on. Additionally, it has been observed that the negative effects of barley on broiler chick growth are not evident in older laying hens, meaning that barley can indeed have a place in poultry rations (Classen et al., 1988b).

Dunstan (1973) reported findings that replacing mixed-grain diets of wheat and oats with barley in laying hens did not significantly affect the number of eggs produced or hen body weights. They did, however, note that the feed consumption was greater for the barley diets as opposed to the mixed-grain diets used for the experiments. This was adjudged to be due to the lower metabolizable energy of the barley diets, which were estimated to be 91% of the mixed-grain diet.

The use of hulless instead of hulled barley in chicken feed rations has been shown to have significant advantages. Removal of the hull results in lower crude fibre and higher levels of valuable nutrients, including crude protein. Hulless barley has been shown as comparable to wheat in the rations of egg laying hens, and contributes to the production of larger eggs and increased body weight over the use of conventional hulled barley (Classen et al., 1988b). In broiler chicks, hulless barley led to higher weight gain and lower feed-conversion rates when compared to hulled barley feed, with -glucanase added to both rations (Campbell et al., 1993).

As with pigs, the locking up of phosphorous in barley as phytate presents both a nutritional and waste problem for poultry producers. It seems likely that low-phytate, hulless barley varieties may be useful to the industry. However, as phytases are relatively inexpensive, their use to reduce the need for expensive phosphorous supplements to address both issues has been very popular in Europe and North America. It seems likely that Australian chicken farmers will follow the same strategy, with lower feed costs being of most interest to broiler chicken producers while the egg industry will have a greater motivation to address waste management issues (Hughes and Choct, 1999).

6 Future trends and research opportunities

There seems little doubt that the largest challenge facing future feed barley production will be the need to develop adapted varieties and production systems which address the threats imposed by climate change. To summarise the available literature, it appears that the forecasted changes to the global climate will have significant impacts on both the quantity and quality of the current sources of all feed grains, including barley. This conjecture has been supported by a range of studies using both in-situ and modelling research techniques.

The projected impact of climate change on grain yield is variable, with many areas likely to suffer from production decline (Doltra et al., 2014; Ghahramani and Moore, 2016; Liu et al., 2017; Cammarano et al., 2019; Pirttioja et al., 2019), while some others may benefit from increased yield potential (Dubey and Sharma, 2018; Masud et al., 2018). A few production areas may see little change in yield and may even benefit from needing to apply less N fertiliser to barley crops (Holden and Brereton, 2006).

The key factor implicated in causing climate change is the increase of atmospheric CO_2. This is a complicating factor in trying to predict the impact of our changing climate as, apart from the impact of changing weather patterns on crop production, yield and quality, higher levels of atmospheric CO_2 are anticipated to affect all aspects of plant growth. As early as the 1990s, researchers were beginning to investigate the potential impact of higher CO_2 levels. In C_3 cereal species such as barley, wheat and rice, increases in atmospheric CO_2 have been shown to result in increased plant growth and grain yield. This is due to the greater sensitivity of the C_3 biological mechanisms to CO_2, as compared to the C_4 species, such as sorghum and maize, which have a tenfold greater CO_2 assimilation rate with low water demand and a better CO_2 capture system (Hashiguchi et al., 2010). A

1974 controlled environment pot study of barley and wheat found that barley responded to increased CO_2 by raising grain yield, largely through increased grain number per plant (Thompson and Woodward, 1994). However, grain nitrogen concentration was significantly reduced as the atmospheric CO_2 was increased. Significantly though, the overall loss of grain nitrogen was proportionately greater than the increase in yield. Another CO_2 study including two spring barley cultivars resulted in lower grain nitrogen, magnesium and calcium concentrations, and lower grain protein concentration with some variation observed in the relative proportion of amino acids between the ambient and increased CO_2 treatments (Manderscheid et al., 1995). Erbs et al. (2010) demonstrated significant reductions in crude protein concentration and sulphur in grain from a winter barley variety grown under elevated CO_2 levels. The data from the study indicated that at least part of the effect was caused by dilution due to increased starch content under the high CO_2 growth conditions.

These trends have been confirmed in a meta-analysis of data from 228 studies in barley, rice, wheat, soybean and potato, which indicated that reduction of grain nitrogen and protein is highly likely to occur across crop species with the increasing atmospheric CO_2 (Taub et al., 2008). In barley, wheat and rice, the impact on grain protein concentration of elevated CO_2 (540–958 µmol mol^{-1}) was a production of 10–15% less protein versus the concentration obtained under the control ambient CO_2 (315–400 µmol mol^{-1}) conditions. Higher rates of N fertiliser can be used to maintain the protein content of cereals under higher atmospheric CO_2 levels (Kimball et al., 2001), however, may be neither economical nor practical under the broad-scale dryland production systems used for most cereals.

Hogy and Fangmeier (2008) identified statistically significant CO_2-induced effects on the grain content of essential amino acids such as threonine, valine, isoleucine, leucine and phenylalanine. The relative globulin and B-hordein protein fractions of barley have been demonstrated to be significantly increased due to an elevated atmospheric CO_2 concentration (Wroblewitz et al., 2014). A meta-analysis which included wheat, rice and corn but not barley indicated that zinc (Zn) and iron (Fe) concentrations also decrease in grain under elevated atmospheric CO_2 levels, noting that both crop hydration and nitrogen fertiliser modified the crop responses (Al-Hadeethi et al., 2019).

However, looking at elevated CO_2 alone presents an unrealistic view of the potential impacts of climate change on grain crop production and quality. Increased temperatures and changed rainfall patterns will also have a significant impact on crop performance and grain quality (Högy et al., 2013), and interactions of all three environmental attributes will come into play in overall crop performance. While higher temperature stress and moisture stress have negative effects on vegetative growth and grain filling, higher CO_2 levels appear to have positive effects on the same traits (Jagadish et al., 2014). However, all three environmental factors have negative effects on grain physical and nutritional quality. A key question, therefore, becomes whether the positive impacts of higher CO_2 on vegetative growth and grain fill are enough to overcome the negative growth effects of heat and moisture stresses, as well as the reduction of grain nutritional quality.

In studies with multiple barley genotypes, the combination of elevated CO_2 and higher temperature was found to have a profound impact on grain yield and protein levels (Ingvordsen et al., 2015; Ingvordsen et al., 2016). While the elevated CO_2 decreased grain protein concentration by 5%, elevated temperatures resulted in a 29% increase in protein content. The net effect of the two treatments applied in combination was an increase of 8% protein content across the full set of genotypes. However, the combined treatments also resulted in a significant yield reduction, with the overall impact of a 23% reduction in harvestable protein. Given the significant effects of temperature and moisture stress on grain size (Eagles et al., 1995), it would seem likely that the overall impact of climate change on the global barley crop will be a large increase in the rejection of grain from the malting receival specifications. This should lead to a proportionate increase in bulk feed barley, which may help to offset some of the tonnages lost from droughted growing regions. However, the feed value of the bulk grain is likely to be less than the current situation, as a direct consequence of the likelihood of reduced grain protein content due to the higher atmospheric CO_2.

There is evidence that concerted research and development into adaptation of farming systems, changes to crop management and plant breeding can offset at least some of the impacts of increased temperatures, rainfall variability and other notable factors (Nendel et al., 2014; Liu et al., 2017; Ghahramani and Bowran, 2018). However, much more research is required.

Barley seems likely to be more amenable to breeding efforts to minimise the impact of climate change on bulk grain production, because its proven broad adaptability when compared to other cereal crops means that the necessary genetic variation is likely to exist to allow plant breeders to develop new varieties. Barley is grown in climatic zones which range from cold through to subtropical. We know that grain size in barley is highly heritable (Fox et al., 2006). Relatively untapped traits in barley such as stay-green (Gous et al., 2013) and root architecture (Abdel-Ghani et al., 2019) are also likely to assist the development of new varieties with improved adaptation to yielding as well as possible under environmental stress.

It seems highly likely that the issue of reduced grain nutritional quality because of climate change will be more difficult to address than plant adaptation. Breeders are in a position where they have access to many modern tools which can be used to speed up and refine the selection cycles to address the multiple stressors (Jagadish et al., 2014). These include the availability of multiple finely mapped QTLs for stress tolerances, technology platforms for high-throughput molecular marker analysis and steadily improving understanding of the genome, proteome and critical developmental pathways. Much of the investigations to date have focussed on individual stressors; however, to successfully address the potential impacts of climate change there will need to be more work conducted on the understanding of, and selection for, tolerance to multiple abiotic stresses.

One scientific discipline which appears to offer substantial promise in understanding plant physiological requirements to overcome the plant growth and grain-quality challenges posed by climate change is that of proteomics.

Proteomics is the study of the full range of proteins extracted from specific tissues, structures or organs and is becoming a powerful tool to understand and enable biotechnological improvement of crops. This includes developing the abilities to understand and specifically modify mechanisms of plant growth and development, responses to biotic and abiotic stresses and processes of seed or fruit development (Tan et al., 2017). Used in combination with fieldwork, the other 'omics' and molecular marker technologies, breeders will be able to direct their selection processes towards the enhanced or reduced expression of particular proteins at specific times in plant growth and development. For example, proteomic studies of leaf senescence and grain development rates will hopefully lead to the selection of cereal varieties which are more tolerant to heat and moisture stresses occurring during grain fill, and thereby, reduce yield loss from heat wave conditions or in warmer environments (Dupont and Altenbach, 2003). To be of most use to plant breeders, the proteome information should to be connected to other useful genetic tools. Such work has commenced, for example, the proteome and molecular marker maps for barley chromosomes 1H, 2H, 3H, 5H and 7H developed and published by Finnie et al. (2009).

In barley, proteomic studies are likely to create opportunities to further improve crop adaptation in response to climate change and for improving grain-quality attributes for malting, feed and food purposes. A substantial amount of work has been done to date and can be tracked down in the literature. As a starting point related to grain quality and the possibilities for use of proteomics in variety improvement, excellent reviews of cereal grain proteomics have been written by Finnie and Svensson, 2009; Finnie et al. (2011) and Labuschagne (2018).

Aqueous extracts from barley indicate the presence of at least 1000 individual proteins in developed grains, with the larger proportion in both concentration and number being found in the aleurone layer and embryo (Finnie and Svensson, 2003). The starchy endosperm contains less than 50% of the grain protein content by weight, despite being the largest component of the grain. Examination of the endosperm proteome indicates that starch synthesis enzymes are found throughout the internal structure of the starch granules, while hordein and serpin storage proteins and two non-starch synthesising enzymes are located on the granule surfaces (Borén et al., 2004). The aleurone and seedling proteomes appear to contain many of the same proteins as the endosperm, plus many others which include the enzymes involved in germination and endosperm breakdown (Finnie and Svensson, 2003; Finnie et al., 2011).

Most proteome studies of barley grain and related to barley end-use quality have been looking at factors which contribute to the making and quality of beer (Colgrave et al., 2013; Lastovicková and Bobálová, 2012; Iimure and Sato, 2013.) In a study including 18 feed and malting cultivars, Finnie et al. (2009) found that the subset of malting cultivars clustered together more closely through a pattern analysis of their proteomes than by a similar analysis of their SSR marker profiles. This supported the selection history of malting varieties by stringent quality assessment in laboratory and processing situations, and indicated that selection by proteome may be a more effective tool to develop high-quality barley varieties. Notably, the feed variety proteomes in the study did not cluster in the same manner as the malting varieties, a likely consequence of the less selective processes and lower emphasis on quality used to develop 'feed' varieties, as discussed earlier in this chapter. However, the clustering of the malting varieties was an indication that proteome selection for feed barley quality characteristics could be a highly effective way to improve the grains nutritional content.

Proteome studies have identified grain quality–related pathways which can be enhanced and/or selected through the DNA manipulation and tracking technologies now available to plant breeders. Examples include grain lysine content (Forsyth et al., 2005), chitinase (for protection from grain attacking pathogens) heat shock and putative nitrogen accumulation-related proteins (Finnie et al., 2004). Kaspar-Shoenefeld et al. (2016) conducted a major study into the proteomic systems in developing barley grain and identified and described 226 proteins expressed during grain fill. Protein activity was at its greatest in the earlier stages of grain development, especially from cellularisation through to the switch to the storage phase or during storage product accumulation. Transgenic C-hordein antisense barley lines were developed and found in a proteomic study to possess an improved balance of beneficial amino acids, as a result of alterations in the non-storage protein profiles in the grain (Schmidt et al., 2016). In particular, the expression of lysine-rich proteins was found to have been upregulated. The increase in lysine was not the product of a single change, but rather a consequence of multiple changes to production of storage and non-storage proteins as well as changes in the lysine mechanism itself. Further research is certainly warranted.

Another topic under investigation through proteomic research is in the identification, monitoring and regulation of toxins in food and feed products (Giacometti et al., 2013). Current and future research opportunities include both detection and monitoring of toxins, and the potential to understand and manipulate the relationship between host and pathogen through which the toxins are produced and enter the grain or other feed material. There are several host pathogen relationships in barley, for example *Fusarium graminearum*, which would warrant investigation to help protect humans and animals from fungal toxins which can be present in or on the grain.

The genetic diversity of barley features many genes and attributes which can be exploited for improving the health, performance and productivity of all grain-fed animal and bird species. Conventional plant breeding methods have already produced successful varieties with a range of useful feed traits, and with the vast amount of information being gathered through the application of new technology platforms, significant additional improvements are highly achievable. Genetic gain is maximised when rapid and convenient genotype testing is available to the plant breeders to tilt the selection probabilities in their favour. This requires knowledge of which traits are the most efficient in both improving performance and heritability. Therefore, there is considerable room for researchers to improve the understanding of the interaction of barley with animal and bird digestive systems and identify additional traits or specifications which can be used to optimise feed performance. Animal proteomics will no doubt feature highly in the research undertaken to explore these possibilities. Alongside the trait discovery phase of this research, is the need for development of rapid, inexpensive and high-volume diagnostic tests to select the best genotypes for advancement in the breeding programmes.

There is also potential to further expand the genetic base in barley. Investigation of the world barley collections and wild populations is ongoing. Mutation, whether spontaneous or induced, has proven to be highly effective in developing new genetic variation for feed-quality traits. Finally, genetic engineering technologies are also likely to contribute significantly to the protein, compositional quality and adaptability of future barley varieties (Forsyth et al., 2005; Shewry, 2007), especially as the demand for feed grain production under increasing environmental and resource pressures continues.

However, very little of this work will be able to proceed without changes in the funding paradigms for barley research and the overall perceptions of feed barley as a bulk commodity created from downgraded malting barley. Significant changes in the approach of both the livestock and grain industries will be essential, in order to achieve a realisation that the best result for all will be to develop a system where both the production gains and profits are shared by all producers and industries from sowing to plate. Increasing pressures from population growth, climate change and the demands for food security are likely to be key drivers to stimulate the required changes to fall into place.

7 Conclusion

This chapter has explored the global barley markets and argued that feed barley should be viewed as comprising a series of different and potentially valuable categories, rather than the current perception as feed being a bulk commodity comprised of failed malting barley. Past and current opportunities for developing and enhancing different styles of barley have been explored, along with traits suited to the production of different species including ruminants, monogastric mammals, poultry and fish. Future challenges such as climate change have been explored, together with the potential of more recent technologies such as proteomics to help address the developing issues.

While primarily exploring the opportunities for optimising the use of barley as a feed through processing and genetic improvement, we have aimed to show that the ultimate solution to get the best feed performance from the crop is to take a higher-level viewpoint. To continue to succeed in the future world, the barley production and utilisation industries must seek to improve the whole system from crop production and stability, through grain handling, processing and ration formulation through to the enhancement of animal management systems.

8 Where to look for further information

There is quite a reasonable amount of additional information available to the researcher with an interest in feed barley. Feed barley specifications can be obtained from the relevant grain-marketing bodies in the various jurisdictions where barley is grown and/ or used. Barley research and development is ongoing as a worldwide activity. Topics such as barley genetics, production agronomics, malting and brewing quality, and, the various 'omics are studied and published around the world. Notable output on specific quality attributes of feed barley has mostly been produced over the past decade from Australia, the United States of America and Canada.

9 References

Abdel-Ghani, A. H., Sharma, R., Wabila, C., Dhanagond, S., Owais, S. J., Duwayri, M. A., Al-Dalain, S. A., Klukas, C., Chen, D., Lübberstedt, T., von Wirén, N., Graner, A., Kilian, B. and Neumann, K. (2019) Genome-wide association mapping in a diverse spring barley collection reveals the presence of QTL hotspots and candidate genes for root and shoot architecture traits at seedling stage. BMC Plant Biol. 19:216. https://bmcplantbiol.biomedcentral.com/track/pdf/10.1186/s12 870-019-1828-5.

Abdel-Haleem, H., Bowman, J., Kanazin V et al. (2010b) Quantitative trait loci of dry matter digestibility and particle size traits in two-rowed 9 six-rowed barley population. Euphytica. 172:419–33.

Abdel-Haleem, H., Bowman, J., Giroux, M., Kanazin, V., Talbert, H., Surber, L. and Blake, T. (2010a) Quantitative trait loci of acid detergent fiber and grain chemical composition in hulled × hull-less barley population. Euphytica. 172:405–18.

Abdel-Haleem, H., Bowman, J. G. P., Surber, L. and Blake, T. (2012) Variation in feed quality traits for beef cattle in Steptoe × Morex barley population. Mol. Breed. 29:503–14.

Aherne, F. X. (1990) Barley: Hulless. In Nontraditional Feed Sources for Use in Swine Production. Ed. Thacker, P. A. and Kirkwood, R. N., Butterworth, Toronto, pp. 23–31.

Al-Hadeethi, I., Seneweera, S., Li, Y., Odhafa, A. K. H., Al-Hadeethi, H. and Lam, S. K. (2019) Assessment of grain quality in terms of functional group response to elevated [CO_2], water, and nitrogen using a meta-analysis: Grain protein, zinc, and iron under future climate. *Ecol. Evol.* 9:7425–37.

Anderson, J. O., Wagstaff, N. K. and Dobson, O. C. (1960) Value of barley and hulless barley in rations for laying hens. *Poult. Sci.* 39:1230 (Abstr.).

Barr, A. R., Jefferies, S. P., Broughton, S., Chalmers, K. J., Kretschmer, J. M., Boyd, W. J. R., Collins, H. M., Roumeliotis, S., Logue, S. J., Coventry, S. J., Moody, D. B., Read, B. J., Poulsen, D., Lance, R. C. M., Platz, G. J., Park, R. F., Panozzo, J. F., Karakousis, A., Lim, P., Verbyla, A. P. and Eckermann, P. J. (2003a) Mapping and QTL analysis of the barley population Alexis × Sloop. *Aust. J. Agric. Res.* 54:1117–23.

Barr, A. R., Karakousis, A., Lance, R. C. M., Logue, S. J., Manning, S., Chalmers, K. J., Kretschmer, J. M., Boyd, W. J. R., Collins, H. M., Roumeliotis, S., Coventry, S. J., Moody, D. B., Read, B. J., Poulsen, D., Li, C. D., Platz, G. J., Inkerman, P. A., Panozzo, J. F., Cullis, B. R., Smith, A. B., Lim, P. and Langridge, P. (2003b) Mapping and QTL analysis of the barley population Chebec × Harrington. *Aust. J. Agric. Res.* 54:1125–30.

Becker, J. and Heun, M. (1995) Barley microsatellites: Allele variation and mapping. *Plant. Mol. Biol.* 27:835–45.

Bedford, M. R. and Morgan, A. J. (1996) The use of enzymes in poultry diets. *World's Poultry Science Journal.* 52:61–8.

Bhatty, R. S., Berdahl, J. D. and Cristison, G. I. (1975) Chemical composition and digestible energy of barley. *Can. J. Anim. Sci.* 55:759–64.

Bhatty, R. S., Christison, G. I. and Rossnagel, B. G. (1979) Energy and protein digestibilities of hulled and hulless barley determined by swine-feeding. *Can. J. Anim. Sci.* 59:585–8.

Black, J. (2018) Premium Grains for Livestock Program, Component 1: Co-ordination, An overview of outcomes from PGLP 1 & 2, Final Report, September 2008. https://ses.library.usyd.edu.au/bit stream/2123/5441/1/PGLPfinalreport.pdf

Blake, T., Bowman, J. P. G., Hensleigh, P., Kushnak, G. et al. (2002) Registration of 'Valier' barley. *Crop. Sci.* 42:1748.

Blake, T., Blake, V. C., Bowman, J. P. G and Abdel-Haleem, H. (2011) Barley feed uses and quality improvement. In *Barley: Production Improvement and Uses*. Ed. Ullrich, S. E., Wiley-Blackwell, pp. 522–31.

Bleidere, M. and Gaile, Z. (2012) Grain quality traits important in feed barley. *Proc. Latvian Acad. Sci., Section B.* 66: No. 1/2.

Borén, M., Larsson, H., Falk, A. and Jansson, C. (2004) The barley starch granule proteome — internalized granule polypeptides of the mature endosperm. *Plant Sci. J.* 166:617–26.

Boss, D. L., Bowman, J. G. P., Surber, L. M. M. Anderson, D. C. and Blake, T. K. (1999) Improving the feed value of barley for finishing steers. *Proc. West. Sec. Am. Soc. Anim. Sci.* 50:293–6.

Bowland, J. P. (1974) Comparison of several wheat cultivars and a barley cultivar in diets for young pigs. *Can. J. Anim. Sci.* 54:629–38.

Bowman, J. G. P., Blake, T. K., Surber, L. M. M., Habernicht, D. K. and Bockelman, H. (2001) Feed-quality variation in the barley core collection of the USDA National Small Grains Collection. *Crop. Sci.* 41:863–70.

Burger, W. C. and Laberge, D. E. (1985) Malting and brewing quality. In *Barley*. Ed. Rasmussen, D. C. American Society of Agronomy: Madison WI, pp. 367–401.

Burnett, G. S. (1966) Studies of viscosity as the probable factor involved in the improvement of certain barleys for chickens by enzyme supplementation. *Brit. Poult. Sci.* 7:55–75.

Cakir, M., Poulsen, D., Galwey, N. W., Ablett, G. A., Chalmers, K. J., Platz, G. J., Park, R. F., Lance, R. C. M., Panozzo, J. F., Read, B. J., Moody, D. B., Barr, A. R., Johnston, P., Li, C. D., Boyd, W. J. R.,

Grime, C. R., Appels, R., Jones, M. G. K. and Langridge, P. (2003) Mapping and QTL analysis of the barley population Tallon × Kaputar. *Aust. J. Agric. Res.* 54:1155–62.

Cammarano, D., Ceccarelli, S., Grando, S., Romagosa, I., Benbelkacem, A., Akar, T., Al-Yassin, A., Pecchioni, N., Francia, E. and Ronga, D. (2019) The impact of climate change on barley yield in the Mediterranean basin. *Eur. J. Agron.* 106:1–11.

Campbell, G. L., Rossnagel, B. G. and Bhatty, R. (1993) Evaluation of hull-less barley genotypes varying in extract viscosity in broiler chick diets. *Anim. Feed Sci. Technol.* 41:191–7.

Clarke, L. C., Sweeney, T., Curley, E., Gath, V., Duffy, S. K., Vigors, S., Rajauria, G. and O'Doherty, J. V. (2018) Effect of -glucanase and -xylanase enzyme supplemented barley diets on digestibility, growth performance and expression of intestinal nutrient transporter genes in finisher pigs. *Anim. Feed Sci. Technol.* 238:98–110.

Classen, H. L., Campbell, G. J., Rossnagel, B. G., Bhatty, R. S. and Reichert, R. D. (1985) Studies on the use of hulless barley in chick diets: Deleterious effects and methods of alleviation. *Can. J. Anim. Sci.* 65:725–33.

Classen, H. L., Campbell, G. L. and Grootwassink, J. W. D. (1988a) Improved feeding value of Saskatchewan-grown barley for broiler chickens with dietary enzyme supplementation. *Can. J. Anim. Sci.* 68:1253–9.

Classen, H. L., Campbell, G. L., Rossnagel, B. G. and Bhatty, R. S. (1988b) Evaluation of hulless barley as replacement for wheat or conventional barley in laying hen diets. *Can. J. Anim. Sci.* 68:1261–6.

Colgrave, M. L., Goswami, H., Howitt, C. A. and Tanner, G. J. (2013) Proteomics as a tool to understand the complexity of beer. *Food. Res. Int.* 54: 1001–12.

Coventry, S. J., Barr, A. R., Eglinton, J. K. and McDonald, G. K. (2003) The determinants and genome locations influencing grain weight and size in barley (Hordeum vulgare L.). *Aust. J. Agric. Res.* 54:1103–15.

Darroch, C. S., Aherne, F. X., Helm, J., Sauer, W. C. and Jaikaran, S. (1996) Effects of dietary level of barley hulls and fiber type on protein and energy digestibilities of Condor hulless barley in growing swine. *Anim. Feed Sci. Technol.* 61:173–82.

Dixon, R. M. and Stockdale, C. R. (1999) Associative effects between forages and grains: Consequences for feed utilisation. *Aust J. Agric. Res.* 50:757–73.

Doltra, J., Laegdsmand, M. and Olesen, J. E. (2014) Impacts of projected climate change on productivity and nitrogen leaching of crop rotations in arable and pig farming systems in Denmark. *J. Agric. Sci.* 152:75–92.

Du, L., Yu, P., Rossnagel, B. G., Christensen, D. A. and McKinnon, J. J. (2009) Physiological characteristics, hydroxycinnamic acids (ferulic acid, p-coumaric acid) and their ratio, and in situ biodegradability: Comparison of genotypic differences among six barley varieties. *J. Agric. Food. Chem.* 57:4777–83.

Dubey, S. K. and Sharma, D. (2018) Assessment of climate change impact on yield of major crops in the Banas River Basin, India. *Sci. Total. Environ.* 635:10–19.

Dunstan, E. A. (1973) The performance of laying hens on diets using barley as the main energy source. *Aust. J. Exp Agr. Anim. Husb.* 13:251–6.

Dupont, F. M. and Altenbach, S. B. (2003) Molecular and biochemical impacts of environmental factors on wheat grain development and protein synthesis. *J. Cereal Sci.* 38:133–46.

Eagles, H. A., Bedggood, A. G., Panozza, J. F. and Martin, P. J. (1995) Cultivar and environmental effects on malting quality in barley. *Aust. J. Agric. Res.* 46:831–44.

Emebiri, L. C., Moody, D. B., Panozzo, J. F., Chalmers, K. J., Kretschmer, J. M. and Ablett, G. A. (2003) Identification of QTLs associated with variations in grain protein concentration in two-row barley. *Aust. J. Agric. Res.* 54:1211–21.

Erbs, M., Manderscheid, R., Jansen, G., Seddig, S., Pacholski, A. and Weigel, H. J. (2010) Effects of free-air CO_2 enrichment and nitrogen supply on grain quality parameters and elemental composition of wheat and barley grown in a crop rotation. *Agr. Ecosyst. Environ.* 136:59–68.

Evers, A. D., Blakeney, A. B. and O'Brien, L. (1999) Cereal structure and composition. *Aust. J. Agric. Res.* 50:629–50.

Farrell, D. J. (1999) In vivo and in vitro techniques for the assessment of the energy content of feed grains for poultry: A review. *Aust. J. Agric. Res.* 50:881–8.

Fedak, G., Tsuchiya, T. and Helgason, S. B. (1972) Use of monotelotrisomics for linkage mapping in barley. *Can. J. Genet. Cytol.* 14:949–57.

Finnie, C. and Svensson, B. (2003) Feasibility study of a tissue-specific approach to barley proteome analysis: Aleurone layer, endosperm, embryo and single seeds. *J. Cereal. Sci.* 38:217–27.

Finnie, C. and Svensson, B. (2009) Barley seed proteomics from spots to structures. *J. Proteom.* 72:315–24.

Finnie, C., Steenholdt, T., Noguera, O. R., Knudsen, S., Larsen, J., Brinch-Pedersen, H., Holm, P. B., Olsen, O. and Svensson, B. (2004) Environmental and transgene expression effects on the barley seed proteome. *Phytochemistry.* 65:1619–27.

Finnie, C., Bagge, M., Steenholdt, T., Østergaard, O., Bak-Jensen, K. S., Backes, G., Jensen, A., Giese, H., Larsen, J., Roepstorff, P. and Svensson, B. (2009) Integration of the barley genetic and seed proteome maps for chromosome 1H, 2H, 3H, 5H and 7H. *Funct. Integr. Genomics.* 9:135–43.

Finnie, C., Sultan, A. and Grasser, K. D. (2011) From protein catalogues towards targeted proteomics approaches in cereal grains. *Phytochemistry.* 72:1145–53.

Forsyth, J. L., Beaudoin, F., Halford, N. G., Sessions, R. B., Clarke, A. R. and Shewry, P. R. (2005) Design, expression and characterisation of lysine-rich forms of the barley seed protein CI-2. *Biochimica et Biophysica Acta.* 1747:221–7.

Fox, G. P., Kelly, A., Poulsen, D., Inkerman, A. and Henry, R. (2006) Selecting for increased barley grain size. *J. Cereal. Sci.* 43:198–208.

Fox, G. P., Nguyen, L., Bowman, J., Poulsen, D., Inkerman, A. and Henry, R. J. (2007) Relationship between hardness genes and quality in barley (Hordeum vulgare). *J. Inst. Brew.* 113:87–95.

Fox, G. P., Bowman, J., Kelly, A., Inkerman, A., Poulsen, D. and Henry, R. (2008) Assessing for genetic and environmental effects on ruminant feed quality in barley (*Hordeum vulgare*). *Euphytica.* 163:249–57.

Fox, G., Kelly, A., Bowman, J., Inkerman, A., Poulsen, D. and Henry, R. (2009) Is malting barley better feed for cattle than feed barley? *J. Inst. Brew.* 115:95–104.

Fox, G., Borgogne, M. G., Flinn, P. and Poulsen, D. (2011) Genetic and environmental analysis of NIR feed quality predictions on genotypes of barley (Hordeum vulgare L.) *Field. Crops. Res.* 20:380–6.

Franckowiak, J. (2011) Barley Breeding Australia - Northern Node DAQ00110. Final Report. http://era .daf.qld.gov.au/id/eprint/3003/1/GRDC_final_report_DAQ00110.pdf.

Friedt, W. (2011) Barley Breeding History, Progress, Objectives, and Technology – Europe. In Barley feed uses and quality improvement. In *Barley: Production Improvement and Uses.* Ed. Ullrich, S. E. Wiley-Blackwell, pp. 160–71.

Gabert, V. M., Brunsgaard, G., Eggum, B. O. and Jensen, J. (1995) Protein quality and digestibility of new high-lysine barley varieties in growing rats. *Plant. Food. Hum. Nutr.* 48:169–79.

Gabert, V. M., Jorgensen, H., Brunsgaard, G., Eggum, B. O. and Jensen, J. (1996) The nutritional value of new high-lysine barley varieties determined with rats and young pigs. *Can. J. Anim. Sci.* 76:443–50.

Ghahramani, A. and Bowran, D. (2018) Transformative and systemic climate change adaptations in mixed crop-livestock farming systems. *Agric. Sys.* 164:236–51.

Ghahramani, A. and Moore, A. D. (2016) Impact of climate changes on existing crop-livestock farming systems. *Agric. Sys.* 146:142–55.

Giacometti, J., Tomljanovi , A. B. and Josi , D. (2013) Application of proteomics and metabolomics for investigation of food toxins. *Food. Res. Int.* 54:1042–51.

Godfrey, S. I., Boyce, M. D., Rowe, J. B. and Speijers, E. J. (1992) Changes within the digestive tract of sheep following engorgement with barley. *Aust. J. Agric. Res.* 44:1093–101.

Gous, P. W., Martin, A., Lawson, W., Kelly, A., Fox, G. P. and Sutherland, M. W. (2012a) QTL associated with barley (Hordeum vulgare) feed quality traits measured through in situ digestion. *Euphytica.* 185:37–45.

Gous, P. W., Martin, A., Lawson, W., Kelly, A., Fox, G. P. and Sutherland, M. W. (2012b) Correlation between NIRS generated and chemically measured feed quality data in barley (Hordeum vulgare), and potential use in QTL analysis identification. *Euphytica.* 188:325–32.

Gous, P. W., Hasjim, J., Franckowiak, J., Fox, G. P. and Gilbert, R. G. (2013) Barley genotype expressing "stay-green"-like characteristics maintains starch quality of the grain during water stress condition. *J. Cereal. Sci.* 58:414–19.

Graner, A., Jahoor, A., Schondelmaier, J., Siedler, H., Pillen, K., Fischbeck, G., Wenzel, G. and Herrmann, R. G. (1991) Construction of an RFLP map of barley. *Theor. Appl. Genet.* 83:250–6.

Gray, R., Nagy, C. and Guzel, A. (2012) Returns to Research, Western Grains Research Foundation, Wheat and Barley Varietal Development, 31 October 2012. http://westerngrains.com/wp-co ntent/uploads/2017/01/Final-WGRF-ROR-STUDY2.pdf.

Grimson, R. E., Weisenburger, R. D., Basarab, J. A. and Stilborn, R. P. (1987) Effects of barley volume-weight and processing method on feedlot performance of finishing steers. *Can. J. Anim. Sci.* 67:43–53.

Hart, K. J., Rossnagel, B. G. and Yu, P. (2008) Chemical characteristics and in-situ ruminal parameters of barley for cattle: Comparison of the malting cultivar AC Metcalfe and five feed cultivars. *Can. J. Anim. Sci.* 88:711–9.

Hashiguchi, A., Ahsan, N. and Komatsu, S. (2010) Proteomics application of crops in the context of climatic changes. *Food. Res. Int.* 43:1803–13.

Hayes, P. M., Liu, B. H., Knapp, S. J., Chen, F., Jones, B., Blake, T., Franckowiak, J., Rasmusson, D., Sorrells, M., Ullrich, S. E., Wesenberg, D. and Kleinhofs, A. (1993) Quantitative trait locus effects and environmental interaction in a sample of North American barley germ plasm. *Theor. Appl. Genet.* 87:392–401.

Heun, M., Kennedy, A. E., Anderson, J. A., Lapitan, N. L. V., Sorrells, M. E. and Tanksley, S. D. (1991) Construction of a restriction fragment length polymorphism map for barley (*Hordeum vulgare*). *Genome.* 34:437–47.

Hogan, J. P. and Flinn, P. C. (1999) An assessment by *in vivo* methods of grain quality for ruminants. *Aust J. Agric. Res.* 50:843–54.

Högy, P. and Fangmeier, A. (2008) Effects of elevated atmospheric CO_2 on grain quality of wheat. *J. Cereal. Sci.* 48:580–91.

Högy, P., Poll, C., Marhan, S., Kandeler, E. and Fangmeier, A. (2013) Impacts of temperature increase and change in precipitation pattern on crop yield and yield quality of barley. *Food. Chem.* 136:1470–7.

Holden, N. M. and Brereton, A. J. (2006) Adaptation of water and nitrogen management of spring barley and potato as a response to possible climate change in Ireland. *Agric. Water. Manag.* 82:297–17.

Holtshausen, L., Beauchemin, K. A., Schwartzkoph-Genswein, K. S., Gonzalez, L. A., McAllister, T. A. and Gibb, D. J. (2011) Performance, feeding behaviour and rumen pH profile of beef cattle fed corn silage in combination with barley grain, corn or wheat distillers' grain or wheat middlings. *Can. J. Anim. Sci.* 91:703–10.

Hughes, R. J. and Choct, M. (1999) Chemical and physical characteristics of grains related to variability in energy and amino acid availability in poultry. *Aust. J. Agric. Res.* 50:689–01.

Hunt, C. W. (1996) Factors affecting the feeding quality of barley for ruminants. *Anim. Feed. Sci. Technol.* 62:37–48.

Hynd, P. J. and Allden, W. G. (1985) Rumen fermentation pattern, postruminal protein flow and wool growth rate of sheep on a high barley diet. *Aust. J. Agric. Res.* 36:451–60.

Igartua, E., Edney, M., Rossnagel, B. G., Spaner, D., Legge, W. G., Scoles, G. J., Eckstein, P. E., Penner, G. A., Tinkier, N. A., Briggs, K. G., Falk, D. E. and Mather, D. E. (2000) Marker-based selection of QTL affecting grain and malt quality in two-row barley. *Crop. Sci.* 40:1426–33.

Ige, D. V., Kiarie, E., Akinremi, O. O., Rossnagel, B., Flatten, D. and Nyachoti, C. M. (2010) Energy and nutrient digestibility in a hulless low-phytate barley fed to finishing pigs. *Can. J. Anim. Sci.* 90:393–9.

Iimure, T. and Sato, K. (2013) Beer proteomics analysis for beer quality control and malting barley breeding. *Food. Res. Int.* 54:1013–20.

Ingvordsen, C. H., Lyngkjær, M. F., Peltonen-Sainio, P., Mikkelsen, T. N and Jørgensen, R. B. (2015) A 10-days heatwave around flowering superimposed on climate change conditions significantly affects production of 22 barley accessions. *Procedia. Environ. Sci.* 29:160–1.

Ingvordsen, C. H., Gislum, R., Jørgensen, J. R., Mikkelsen, T. N., Stockmarr, A. and Jørgensen, R. B. (2016) Grain protein concentration and harvestable protein under future climate conditions. A study of 108 spring barley accessions. *J. Exp. Bot.* 67:2151–8.

Jacobs, J. L., Morris, R. J. and Zorilla-Rios, J. (1995) Effect of ensiling whole barley grain with pasture on silage quality and effluent production and the performance of growing cattle. *Aust. J. Exp Agric.* 35:731–8.

Jagadish, K. S. V., Kadam, N. N., Xiao, G., Melgar, R. J., Bahuguna, R. N., Quinones, C., Tamilselvan, A. and Prasad, P. V. V. (2014) Agronomic and physiological responses to high temperature, drought, and elevated CO_2. *Adv. Agron.* 127:111–56.

Kaiser, A. G. (1999) Increasing the utilisation of grain when fed whole to ruminants. *Aust. J. Agric. Res.* 50:737–56.

Karakousis, A., Barr, A. R., Kretschmer, J. M., Manning, S., Logue, S. J., Roumeliotis, S., Collins, H. M., Li, C. D., Lance, R. C. M. and Langridge, P. (2003a) Mapping and QTL analysis of the barley population Galleon × Haruna Nijo. *Aust. J Agric. Res.* 54:1131–5.

Karakousis, A., Barr, A. R., Kretschmer, J. M., Manning, S., Jefferies, S. P., Chalmers, K. J., Islam, A. K. M. and Langridge, P. (2003b) Mapping and QTL analysis of the barley population Clipper × Sahara. *Aust. J. Agric. Res.* 54:1137–40.

Karakousis, A., Gustafson, J. P., Chalmers, K. J., Barr, A. R. and Langridge, P. (2003c) A consensus map of barley integrating SSR, RFLP and AFLP markers. *Aust J. Agric. Res.* 54:1173–85.

Kaspar-Schoenefeld, S., Merx, K., Jozefowicz, A. M., Hartmann, A., Seiffert, U., Weschke, W., Matros, A. and Mock, H. P. (2016) Label-free proteome profiling reveals developmental-dependent patterns in young barley grains. *J. Proteom.* 143:106–21.

Kenney, P. A. (1986) Productivity of early-weaned lambs fed high-grain diets of wheat, oats or barley with or without lupin grain. *Aust. J. Exp. Agric.* 26:279–84.

Khorasani, G. R., Helm, J. and Kennelly, J. J. (2000) In situ rumen degradation characteristics of sixty cultivars of barley grain. *Can. J. Anim. Sci.* 80:691–701.

Kimball, B. A., Morris, C. F., Pinter Jr, P. J., Wall, G. W., Hunsaker, D. J., Adamsen, F. J., LaMorte, R. L., Leavitt, S. W., Thompson, T. L., Matthias, A. D. and Brooks, T. J. (2001) Elevated CO_2, drought and soil nitrogen effects on wheat grain quality. *New. Phytol.* 150: 295–303.

Kitessa, S., Flinn, P. C. and Irish, G. G. (1999) Comparison of methods used to predict the in vivo digestibility of feeds in ruminants. *Aust. J Agric. Res.* 50:825–41.

Kleinhofs, A., Kilian, A., Saghai Maroof, M. A., Biyashev, R. M., Hayes, P., Chen, F. Q., Lapitan, N., Fenwich, A., Blake, T. K., Kanazin, V., Ananiev, E., Dahleen, L., Kudrna, D., Bollinger, J., Knapp, S. J., Liu, B., Sorrells, M., Heun, M., Franckowiak, J. D., Hoffman, D., Skadsen, R. and Steffenson, B. J. (1993) A molecular, isozyme and morphological map of the barley (*Hordeum vulgare*) genome. *Theor. Appl. Genet.* 86:705–12.

Koenig, K. M. and Beauchemin, K. A. (2011) Optimum extent of barley grain processing and barley silage proportion in feedlot cattle diets: Growth, feed efficiency and faecal characteristics. *Can. J. Anim. Sci.* 91:411–22.

Labuschagne, M. T. (2018) A review of cereal grain proteomics and its potential for sorghum improvement. *J. Cereal. Sci.* 84:151–8.

Larson, S. R., Young, K. A., Cook, A., Blake, T. K. and Raboy, V. (1998) Linkage mapping of two mutations that reduce phytic acid content of barley grain. *Theor. Appl. Genet.* 97: 141–6.

Lastovicková, M. and Bobálová, J. (2012) MS based proteomic approaches for analysis of barley malt. *J. Cereal. Sci.* 56: 519–30.

Lawrence, T. L. J. (1970) High level cereal diets for the growing-finishing pig. IV. Comparison of two slaughter weights (120 and 200lb.) of diets containing high levels of maize, sorghum, wheat and barley. *J. Agr. Sci.* 74:539–48.

Liu, D. L., Zeleke, K. T., Wang, B., Macadam, I., Scott, F. and Martin, R. J. (2017) Crop residue incorporation can mitigate negative climate change impacts on crop yield and improve water use efficiency in a semiarid environment. *Eur. J. Agron.* 85:51–68.

Long, N. R., Jefferies, S. P., Warner, P., Karakousis, A., Kretschmer, J. M., Hunt, C., Lim, P., Eckermann, P. J. and Barr, A. R. (2003) Mapping and QTL analysis of the barley population Mundah × Keel. *Aust. J. Agric. Res.* 54:1163–71.

Lott, J. N., Ockenden, I., Raboy, V. and Batten, G. D. (2000) Phytic acid and phosphorus in crop seeds and fruits: A global estimate. *Seed. Sci. Res.* 10: 11–33.

Manderscheid, R., Bender, J., Jager, H. J. and Weigel, H. J. (1995) Effects of season long CO_2 enrichment on cereals. II. Nutrient concentrations and grain quality. *Agr. Ecosyst. Environ.* 54:175–85.

Mannion, P. F. (1981) Enzyme supplementation of barley based diets for broiler chickens, *Aust. J. Exp. Agric. Anim. Husb.* 21:296–302.

Masud, M. B., McAllister, T., Cordeiro, M. R. C. and Faramarzi, M. (2018) Modeling future water footprint of barley production in Alberta, Canada: Implications for water use and yields to 2064. *Sci. Total. Environ.* 616–17:208–22.

Mather, D. E, Tinker, N. A., LaBerge, D. E., Edney, M., Jones, B. L., Rossnagel, B. G., Legge, W. G., Briggs, K. G., Irvine, R. B., Falk, D. E. and Kasha, K. J. (1997) Regions of the genome that affect grain and malt quality in a North American two-row barley cross. *Crop. Sci.* 37:544–54.

Merritt, N. R. (1967) A new strain of barley with starch of high amylose content. *J. Inst. Brewing.* 73:583–6.

Mitchall, K. G., Bell, J. M. and Sosulski, F. W. (1976) Digestibility and feeding value of hulless barley for pigs. *Can. J. Anim. Sci.* 56:505–11.

Molina-Cano, J. L, Francesh, M., Perez-Vendrell, A. M., Ramo, T., Voltas, J. and Brufau, J. (1997) Genetic and environmental variation in malting and feed quality of barley. *J. Cereal. Sci.* 25:37–47.

Moughan, P. J. (1999) In vitro techniques for the assessment of the nutritive value of feed grains for pigs: A review. *Aust J. Agric. Res.* 50:871–79.

Munck, L., Karlsson, K. E., Hagberg, A. and Eggum, B. P. (1970) Gene for improved nutritional value in barley seed protein. *Science* 168:985–7.

Nasir, Z., Wang, L. F., Young, M. G., Swift, M. L., Beltranena, E. and Zijlstra, R. T. (2015) The effect of feeding barley on diet nutrient digestibility and growth performance of starter pigs. *Anim. Feed. Sci. Technol.* 210:287–94.

Nendel, C., Kersebaum, K. C., Mirschel, W. and Wenkel, K. O. (2014) Testing farm management options as climate change adaptation strategies using the MONICA model. *Eur. J. Agron.* 52:47–56.

O'Brien, L. (1999) Genotype and environmental effects on feed grain quality. *Aust. J. Agric. Res.* 50:703–19.

Oram, R. N. and Doll, H. (1981) Yield improvement in high lysine barley. *Aus. J. Agric. Res.* 32:425–34.

Pallotta, M. A., Asayama, S., Reinheimer, J. M., Davies, P. A., Barr, A. R., Jefferies, S. P., Chalmers, K. J., Lewis, J., Collins, H. M., Roumeliotis, S., Logue, S. J., Coventry, S. J., Lance, R. C. M., Karakousis, A., Lim, P., Verbyla, A. P. and Eckermann, P. J. (2003) Mapping and QTL analysis of the barley population Amagi Nijo × WI2585. *Aust. J. Agric. Res.* 54:1141–4.

Parsons, J. G. and Price, D. B. (1974) Search for barley (*Hordeum vulgare*) with high lipid content. *Lipids.* 9:804–8.

Perrott, L. A., Strydhorst, S. M., Hall, L. M., Yang, R. C., Pauly, D., Gill, K. S. and Bowness, R. (2018a) Advanced agronomic practices to maximise feed barley yield, quality and standability in Alberta, Canada. I. Responses to plant density, a plant growth regulator, and foliar fungicides. *Agron J.* 110:1447–57.

Perrott, L. A., Strydhorst, S. M., Hall, L. M., Yang, R. C., Pauly, D., Gill, K. S. and Bowness, R. (2018b) Advanced agronomic practices to maximise feed barley yield, quality, and standability in Alberta, Canada. II. Responses to supplemental post-emergence nitrogen. *Agron J.* 110:1458–66.

Peterson, D. S., Harris, D. J., Rayner, C. J., Blakeney, A. B and Choct, M. (1999) Methods for the analysis of premium livestock grains. *Aust. J. Agric. Res.* 50:775–87.

Pierce, J. (2000) Phytase, production and pollution. In *Concepts in Pig Science 2000.* Ed. Lyons, T. P. and Cole, D. J. A. Nottingham University Press, Nottingham, UK, pp. 97–116.

Pirttioja, N., Palosuo, T., Fronzeka, S., Räisänen, J., Rötter, R. P. and Carter, T. R. (2019) Using impact response surfaces to analyse the likelihood of impacts on crop yield under probabilistic climate change. *Agr. Forest. Meteorol.* 264:213–24.

Pointillart, A., Fontaine, N. and Thomasset, M. (1984) Phytate phosphorus utilization and intestinal phytases in pigs fed low phosphorus wheat or corn diets. *Nutr. Rept. Int.* 29: 473–83.

Poulsen, D. (2006) DAQ00038 - Barley Improvement for the GRDC Northern Region. Final Report. https://grdc.com.au/research/reports/report?id=224.

Purchas, R. W. and Lloyd Davies, H. (1974) Carcass and meat quality of Friesian steers fed on either pasture or barley. *Aust. J. Agric Res.* 25:183–92.

Ravindran, V. and Bryden, W. L. (1999) Amino acid availability in poultry – *in vitro* and *in vivo* measurements. *Aust. J. Agric. Res.* 50:889–908.

Read, B. J., Raman, H., McMichael, G., Chalmers, K. J., Ablett, G. A., Platz, G. J., Raman, H., Genger, R. K., Boyd, W. J. R., Li, C. D., Grime, C. R., Park, R. F., Wallwork, H., Prangnell, R. and Lance, R. C. M. (2003) Mapping and QTL analysis of the barley population Sloop × Halcyon. *Aust. J. Agric. Res.* 54:1145–53.

Reynolds, W. K., Hunt, C. W., Eckert, J. W. and Hall, M. H. (1992) Evaluation of the feeding value of barley as affected by variety and location using near infrared reflectance spectroscopy. *Proc. West. Sect. Am. Soc. Anim. Sci.* 43:498–501.

Rossnagel, B. G., Zatorski, T., Arganosa, G. and Beattie, A. D. (2008) Registration of 'CDC Lophy-I' Barley. *J. Plant. Registr.* 2: 169–73.

Rowe, J. B., Choct, M. and Pethick, D. W. (1999) Processing cereal grains for animal feeding. *Aust. J. Agric. Res.* 50:721–36.

Savin, R. and Nicolas, M. E. (1999) Effects of timing of heat stress and drought on growth and quality of barley grains. *Aust. J. Agric. Res.* 50:357–64.

Schmidt, D., Gaziola, S. A., Boaretto, L. F. and Azevedo, R. A. (2016) Proteomic analysis of mature barley grains from C-hordein antisense lines. *Phytochemistry.* 125:14–26.

See, D., Kanazin, V., Kephart, K. and Blake, T. (2002) Mapping genes controlling variation in barley grain protein concentration. *Crop. Sci.* 42:680–5.

Shewry, P. R. (2007) Improving the protein content and composition of cereal grain. *J. Cereal. Sci.* 46:239–50.

Shewry, P. R., Tatham, A. S., Halford, N. G., Barker, J. H. A., Hannappel, U., Gallois, P., Thomas M and Kreis, M. (1994) Opportunities for manipulating the seed protein composition of wheat and barley in order to improve quality. *Transgenic. Res.* 3:3–12.

Sormunen-Cristian, R. (2013) Effect of barley and oats on feed intake, live weight gain and some carcass characteristics of fattening lambs. *Small. Rumin. Res.* 9:22–7.

Sparrow, D. H. B and Doolette, J. B. (1975) Barley. In *Australian Field Crops: Wheat and Other Temperate Cereals*. Ed. Lazenby, A. and Matheson, E. M. Vol. 1. 2nd edn. Angus and Robinson Publishers, Australia.

Surber, L. M. M., Bowman, J. G. P., Blake, T. K., Hinman, D. D., Boss, D. L. and Blackhurst, T. C. (2000) Prediction of barley feed quality for beef cattle from laboratory analyses. *Proc. West. Sec. Am. Soc. Anim. Sci.* 51:454–7.

Tan, B. C., Lim, Y. S. and Lau, S. (2017) Proteomics in commercial crops: An overview. *J. Proteom.* 169:176–88.

Taub, D. R., Miller, B. and Allen, H. (2008) Effects of elevated CO_2 on the protein concentration of food crops: A meta-analysis. *Glob. Chang. Biol.* 14:565–75.

Thacker, P. A., Bell, J. M., Classen, H. L., Campbell, G. L. and Rossnagel, B. G. (1988) The nutritive value of hulless barley for swine. *Anim. Feed Sci. Technol.* 19:191–6.

Thacker, P. A., Rossnagel, B. G. and Raboy, V. (2003) Phosphorus digestibility in low-phytate barley fed to finishing pigs. *Can. J. Anim. Sci.* 83:101–4.

Thompson, G. B. and Woodward, F. I. (1994) Some influences of CO_2 enrichment, nitrogen nutrition and competition on grain yield and quality in spring wheat and barley. *J. Exp. Bot.* 45:937–42.

Toland, P. C. (1976) The digestibility of wheat, barley or oat grain fed wither whole or rolled at restricted levels with hay to steers. *Aust. J. Exp. Agr. Anim. Husb.* 16:71–5.

Valentine, S. C and Wicks, R. B. (1982) The production and composition of milk from dairy cows fed hay and supplemented with either brewers grains or rolled barley grain. *Aust. J. Exp. Agric. Anim. Husb.* 22:155–8.

Van Barneveld, R. J. (1999a) Chemical and physical characteristics of grains related to variability in energy and amino acid availability in pigs: A review. *Aust. J. Agric. Res.* 50:667–87.

Van Barneveld, S. L. (1999b) Chemical and physical characteristics of grains related to variability in energy and amino acid availability in ruminants: A review. *Aust. J. Agric. Res.* 50:651–66.

Van Barneveld, R. J., Nuttall, J. D., Flinn, P. C. and Osborne, B. G. (1999) Near infrared reflectance measurement of the digestible energy content of cereals for growing pigs. *J. Near. Infrared. Spec.* 7, 1–7.

Vasos, E. J., Barr, A. R. and Eglinton, J. K. (2004) Genetic conversion of feed barley varieties to malting types. *Proceedings of the 9th International Barley Genetics Symposium. Czech, June:* 20–6.

White, C. L. and Ashes, J. R. (1999) A review of methods for assessing the protein value of grain fed to ruminants. *Aust. J. Agric. Res.* 50:855–69.

Woyengo, T. A., Akinremi, O. O., Rossnagel, B. G. and Nyachoti, C. M. (2012) Performance and total tract nutrient digestibility of growing pigs fed hulless low phytate barley. *Can. J. Anim. Sci.* 92:505–11.

Wrigley, C. W. (1999) Potential methodologies and strategies for the rapid assessment of feed-grain quality. *Aust. J. Agric. Res.* 50:789–805.

Wroblewitz, S., Hüther, L., Manderscheid, R., Weigel, H. J., Wätzig, H. and Dänicke, S. (2014) Effect of rising atmospheric carbon dioxide concentration on the protein composition of Cereal Grain. *J. Agric. Food. Chem.* 62:6616–25.

Yahaghi, M., Liang, J. B., Balcells, J., Valizadeh, R., Alimon, A. R. and Ho, Y. W. (2012) Effect of replacing barley with corn or sorghum grain on rumen fermentation characteristics and performance of Iranian Baluchi lamb fed high concentrate rations. *Anim. Prod. Sci.* 52:263–8.

Yu, P., Meier, J. A., Christensen, D. A., Rossnagel, B. G. and McKinnon, J. J. (2003) Using the NRC-2001 model and the DVE/OEB system to evaluate nutritive values of Harrington (malting-type) and Valier (feed-type) barley for ruminants. *Anim. Feed Sci. Technol.* 107:45–60.

Chapter 4

Sorghum as a forage and energy crop

Scott Staggenborg and Hui Shen, Chromatin Inc., USA

1 Introduction

Sorghum is an important source of grain and fodder, forage and biomass throughout the world. In the United States, grain sorghum is the prominent crop; however, non-grain or forage sorghum plays an important role as a feedstock globally. Forage sorghum is phenotypically diverse with cultivars that are used primarily for silage, cultivars that may or may not include sudangrass that are primarily used for haying and grazing, and more recently, high-yielding and sweet cultivars that have been positioned for use as renewable feedstocks. Sorghum was quickly identified as one of the most appropriate dedicated energy crops. In addition, having scalable and well-understood production practices makes sorghum more acceptable than many lesser developed perennial grasses. Relative high yields make it an excellent choice over grain-based annuals, and because it is generally not used for human food, it is of less concern in the food-versus-fuel debate (Staggenborg et al., 2008). Sorghum has been used as a renewable feedstock to produce cellulosic ethanol, steam via combustion and methane via anaerobic digestion. Traits such as brown midrib (BMR) and brachytic dwarfism, and heat and drought tolerance make sorghum adaptable to many marginal environments.

2 Forage and biomass sorghum types

The economic yield of sorghum can be defined as grain, forage and biomass, and sugars depending on the end use and the genetic composition of a particular hybrid. This section focuses on three groups: forage and the sweet sorghum hybrids, sorghum × sudangrass

http://dx.doi.org/10.19103/AS.2017.0015.24

(So×Su) and sudangrass (Su×Su). So×Su hybrids are known for high biomass yields and rapid regrowth, and Su×Su hybrids are characterized by rapid early season growth, rapid regrowth, fine stems and lower potential for hydrogen cyanide production, when the plant is damaged, wounded or fed on, compared with forage sorghums and So×Su hybrids.

2.1 Plant height and maturity

These three groups can be further subdivided on the basis of hybrid maturity. Hybrid maturity is controlled by seven flowering genes: *Ma1* through *Ma7*, which influence the duration of growth, or days to maturity (Quinby, 1974; Rooney et al., 2007; Mullet et al., 2010). *Ma1* (Murphy et al., 2011), *Ma3* (Childs et al., 1997; Yang et al., 2014), *Ma5* and *Ma6* (Murphy et al., 2014; Mullet and Rooney, 2013) had been cloned. *Ma2*, *Ma4* and *Ma7* genes are not known, although there are suggested candidate genes (or proposed gene loci) available based on QTL mapping studies (Hart et al., 2001; Rooney and Aydin, 1999). *Ma1* is considered to have large effect on maturity because it has direct role in repressing flowering time. Plants with the *ma1* mutant tend to flower early, and thus produce shorter plants with fewer internode numbers. *Ma1* encodes a pseudoresponse regulator protein 37 (Murphy et al., 2011). It represses the expression of *FT* (*FLOWERING LOCUST*) through enhancing the expression of *CO* (*CONSTANS*) and inhibiting *Ehd1* (*EARLY HEADING DATE 1*) (Doi et al., 2004). *Ma1* has a known allelic series and these mutants were classified into seven groups depending on the mutations *Prr37* gene carries (Murphy et al., 2011; Klein et al., 2015). All these mutants produce photoperiod-insensitive phenotypes under day lengths that are longer than 12 hours.

Manipulation of these genes can result in a forage sorghum, So×Su hybrid, or Su×Su hybrid that either produce a head after some variable vegetative stage (headed) or do not produce a head during a normal growing season in a temperate environment (photoperiod sensitive). Photoperiod-sensitive hybrids are those that require decreasing day length, a particular temperature and a certain photoperiod to trigger floral initiation (Craufurd et al., 1999; Dingkuhn et al., 2008). In temperate environments, this photoperiod response results in floral initiation after a photoperiod of approximately 12 h and 20 min, so panicles are rarely observed or are present late enough to produce no grain. Because these hybrids do not flower until very late in the year, they become very tall and produce very high biomass yields. They are traditionally higher in moisture at harvest because they do not head and trigger a natural dry down process like headed hybrids. From a forage-quality perspective, these hybrids produce no grain and are often considered to have low forage quality.

Sorghum height is mainly controlled by four genes (dwarf): Dw_1, Dw_2, Dw_3 and Dw_4. Genes Dw_1 and Dw_2 are known to affect internode length, Dw_3 affects internode number and Dw_4 appears to affect panicle length (Goud and Vasudev Rao, 1977). Only *Dw1* and *Dw3* genes have been cloned so far. Recent studies suggest *Dw1* gene regulates the length of the internodes. It does not affect leaf size and other plant parts (Klein et al., 2015). The *dw1dw1* can be used to reduce internode length (plant height) without reducing the canopy and leaf development (Klein et al., 2015, Hilley et al., 2016). The *Dw3* gene plays important roles in controlling plant height through regulating total plant height, and flag leaf height and has little impact on panicle exertion (Quinby and Karper 1961; Li et al., 2015). Interestingly, a QTL loci, *qHT7.1*, was found to regulate the interval between flag leaf to apex only (Quinby and Karper 1961). Grain sorghum hybrids have been dwarfed to facilitate mechanized harvest and seldom exceed 2 m in height. Most grain sorghum hybrids developed in the United States are recessive at three height loci (3-dwarf) and their genotypes are generally $dw_1Dw_2dw_3dw_4$ (Quinby and Karper, 1954). One height gene, $dw3$, is known only to reduce internodes on the lower part of the stalk and has little impact on panicle exsertion. From a forage perspective, plant height results in three distinct forage sorghum types: traditional, dual-purpose and brachytic dwarf forage hybrids. It is possible to find brachytic dwarf So×Su hybrids.

2.2 Brown midrib and brachytic dwarf sorghum

Lignin decreases forage digestibility, so efforts to reduce lignin in forage sorghum hybrids have been conducted for decades. BMR mutations (*bmr1* through *bmr19*) were introduced into sorghum by Porter et al. (1978) and Bittinger et al. (1981) via chemical mutagenesis of two grain sorghum lines. Allelism tests of these identified *bmr* mutants result in four allelic groups, *bmr2*, *bmr6*, *bmr12* and *bmr19* (Saballos et al., 2008). Fritz et al. (1990) selected three from this original population that were agronomically acceptable, and they have been the source of three alleles: *bmr6*, *bmr12* and *bmr18*. Since the *bmr19* mutant is not publicly available, and it has limited value for forage and bioenergy applications due to less reduction of lignin content (Saballos et al., 2008), most of the developed *bmr* varieties are based on *bmr6* and *bmr12* mutants (Pedersen et al., 2006a,b,c, 2008). Sorghum *Bmr6* gene encodes cinnamyl alcohol dehydrogenase (Saballos et al., 2009; Sattler et al., 2009), which catalyses the last step in monolignol biosynthesis. Sorghum *Bmr12* gene encodes caffeic acid 3-*O*-methyltransferase (COMT) (Bout and Vermerris, 2003) and sorghum *Bmr2* gene encodes a 4-coumarate coenzyme A ligase (4CL), which catalyses an early step in monolignol biosynthesis (Saballos, et al., 2012). BMR hybrids have noticeable brown pigment in the midrib of the leaves, the stem, the pith and immature panicles (Porter et al., 1978). These hybrids have reduced lignin content and generally higher digestibility, yet they are often low in yield (Porter et al., 1978; Bucholtz et al., 1980; Hanna et al., 1981; Bean and McCollum, 2006; McCuistion et al., 2009, 2010).

Traditional forage sorghums have no recessive dwarf genes, resulting in hybrids that can exceed 4 m in height. While brachytic dwarf forage hybrids have a similar leaf number as traditional forage sorghum hybrids, they have brachytic mutations, which reduce internode length, resulting in a 'stacking' appearance of the leaves. It is believed that because of their higher leaf-to-stem ratios, brachytic dwarf hybrids are more digestible compared with traditional forage sorghum hybrids. Plant height is also important for breeding forage (biomass) sorghum because sorghum plants with increased stature always invite lodging under wind stress. Early BMR hybrids exhibited above average lodging. To overcome this problem, dwarf sorghum mutants were used to breed BMR lines, which will be discussed in the next section. According to Cook (1915), plants with short statues were classified as dwarfs and brachytes. A dwarf is a plant deficient in stature and parts comparing to a normal plant. But a brachytic plant only has a shorter vertical axis, caused by shortening of the internodes, without reduction of other parts such as plant leaf length or size (Cook, 1915). To distinguish these two dwarf types, people sometimes call the brachytic a 'brachytic dwarf'. Plants with longer flowering time, such as photoperiod-sensitive lines, may produce more leaf numbers. However, longer flowering time normally results in taller plant if dwarf genes are not present. As reviewed above, the *dw1* allele can be used to breed brachytic dwarf plant if proper maturity genes are present. The combination of both *ma1ma1dw1dw1* alleles may result in brachytic dwarf plants with less internode numbers. It was reported that *Ma1* and *Dw2* loci are tightly linked on chromosome 6 (Klein et al.,

2008; Multani et al., 2003). The impacts of *ma1ma1dw2dw2dw1dw1* alleles to brachytic dwarf phenotype are not known. Achieving a desirable brachytic dwarf phenotype with larger number of leaves and early maturity through manipulating both the *Ma* genes and *Dw* genes is difficult, as early flowering time will most likely produce fewer leaves due to the developmental transition from vegetative to reproductive stage. A new trait with rapid cell division to produce many, shorter internodes before flowering will benefit early-maturity brachytic dwarf products.

2.3 Water use

Sorghum tolerance to drought and heat stress has been reported by a large number of researchers. Forage sorghum is often compared with corn for improved water-use efficiency and for producing higher yields under water-limiting scenarios (Table 1). Marsalis et al. (2009) reported forage sorghum yields greater or equal to corn silage when the amount of applied irrigation water was reduced to two-thirds of what would have been applied to a fully irrigated corn crop. Howell et al. (2008) reported that water use in BMR sorghum was approximately 73% of that in corn water, as measured by a lysimeter, and Rajan and Maas (2007) reported higher daily water-use rates for corn compared with forage sorghum via eddy covariance and spectral crop coefficient methods in large-scale production fields.

McCuistion et al. (2009) compared the water use and yields of BMR forage sorghum, traditional forage sorghum and photoperiod forage sorghum. Photoperiod forage sorghum had the highest yields and greatest water-use efficiency, 44 kg ha^{-1} dry matter mm^{-1} of water. The water-use efficiency of headed forage sorghums was 21 kg ha^{-1} dry matter mm^{-1} of water. Saeed and El-Nadi (1998) reported water-use efficiencies for forage sorghum ranging from 65 to 86 kg ha^{-1} mm^{-1}, with light, frequent irrigation resulting in the highest water-use efficiencies and dry matter yields. Ottman (2010) reported the greatest water-use efficiency at 75% of evapotranspiration (ET) replacement for forage sorghum. He reported an applied water-use efficiency of 78 kg ha^{-1} mm^{-1}. Hussein and Alva (2014) reported lower water-use efficiency for forage sorghum, 6 kg ha^{-1} mm^{-1}. They did not

Table 1 Silage yields of corn, forage sorghum, and BMR forage sorghum in New Mexico under limited irrigation in 2005 and 2006

	2005	2006	2-yr average
Optimum harvest (60–65% moisture)		Mg ha^{-1}	
Corn	48.7	55.3	51.9
Forage sorghum	61.5	59.0	60.3
BMR-FS	51.6	44.7	48.4
Late harvest (50–60% moisture)			
Corn	57.3	56.3	56.8
Forage sorghum	54.8	57.0	56.1
BMR-FS	46.7	43.5	45.2
SEM	1.1	1.1	1.1

[†] Data from Marsalis et al., 2009; irrigation: 18–20 inches, or about two-thirds of full corn irrigation amounts.

state the cultivar planted, so it is difficult to properly compare the data due to the diversity of forage sorghum types and yield potentials. Narayanan et al. (2013) reported water-use efficiencies ranging from 34 to 76 kg ha^{-1} mm^{-1} from eight public biomass sorghum varieties. Their results illustrate the importance of biomass yield to improve water-use efficiency. Water use ranged from 218 to 256 kg m^{-2} during 2 years, a relatively narrow range compared with yields that ranged from 796 to 1717 g m^{-2}.

3 Forages as animal feed

Forage sorghum, So×Su, and Su×Su hybrids have traditionally been included in rations for ruminant animals. These animals' unique digestive systems are capable of breaking down the complex carbohydrates into simple sugars, which are easily digested. Traditional harvesting and storage methods include grazing, greenchop, dry hay or silage.

3.1 Direct feeding and dry hay methods

Direct-feeding methods, grazing and greenchop require additional management to reduce the potential of negative effects of prussic acid and nitrate poisoning. Prussic acid levels can be reduced by delaying grazing or greenchop harvest until plant heights exceed 45 cm. Prussic acid is concentrated in the leaf tissue, so delaying direct feeding and greenchop harvest until this height is achieved reduces the leaf-to-stem ratio. Nitrates accumulate in grasses that experience growth-reducing stress, drought being the most common. Reductions in dry matter because of stress result in higher concentrations of nitrates, which accumulate in the plant early in the growing season. Nitrate management in a direct-feeding scenario should be managed by testing the forage before feeding.

Sorghum × sudangrass and Su×Su hybrids are the primary hybrids used to produce dry hay. These hybrids can be harvested multiple times during the summer when adequate solar radiation and wind are available to reduce forage moisture content from 80 to 15% (Hill, 1976). When forage sorghum hybrids are harvested for baled hay, additional crimping is required to break the large internodes and aid in the drying process (Hartley et al., 2011).

3.2 Silage

Sorghum silage is predominately produced from forage sorghum hybrids, although photoperiod-sensitive So×Su hybrids are occasionally used in a single-cut system to produce silage. Forage sorghum hybrids can often be divided into three categories: traditional, dual-purpose and brachytic dwarf. The inclusion of BMR genes is common in all three types.

Dual-purpose hybrids are dwarfed forage hybrids that resemble grain sorghum hybrids. Dual-purpose forage sorghums grow to between 1.5 and 2.5 m tall, depending on growing conditions, and will produce a significant amount of grain (Table 2). Lab analyses of silage from these hybrids will always result in high starch levels (~20%) because of the grain in the silage (Table 2). If, however, the plants are not harvested correctly, this starch availability may mean that the lab results are misleading for a nutritionist (e.g. although the lab report may read 25% starch, only 15% may be available to the animal). In order for animals to digest all of this starch, harvest must occur between the milk and soft-dough stages. As the grain

Table 2 Grain and fodder yield as well as compositional analyses of three different forage sorghum types

Variable	Sorghum Partners NK 300	Sorghum Partners SS405	Sorghum Partners 1990
Harvest moisture[†]	62.1	65.2	70.0
Yield (Mg ha⁻¹ @ 65% moisture)[†]	50.0	58.7	69.2
Grain (Mg ha⁻¹)[†]	7.9	2.9	0.0
Crude protein (%)[‡]	9.76 ± 0.54[§]	8.01 ± 0.31	9.54 ± 1.36
Acid detergent fibre (%)[‡]	24.96 ± 0.06	34.63 ± 1.52	33.16 ± 3.61
Neutral detergent fibre (%)[‡]	39.50 ± 1.13	54.43 ± 2.5	54.54 ± 0.62
Total digestible nutrients (%)[‡]	65.24 ± 2.78	65.75 ± 0.35	61.26 ± 4.45
Starch (%)[‡]	33.64 ± 5.15	19.79 ± 9.73	5.93 ± 1.92
Crude fat (%)[‡]	2.25 ± 0.15	2.08 ± 0.52	0.84 ± 0.05
Ash (%)[‡]	6.68 ± 0.76	7.43 ± 0.41	8.65 ± 0.37
Net energy$_{lactation}$ (Mcal/kg)[‡]	0.77 ± 0.43	0.65 ± 0.02	0.67 ± 0.08

[†] From Bean and McCollum 2006.
[‡] Chromatin internal data collected from different samples.
[§] Standard deviation.

matures (proceeds from hard-dough to maturity), it becomes more difficult to digest because of grain hardness. At this point, the grain must be cracked or ground for the starch to be fully digested by the animal.

Full-season headed forage sorghum hybrids are preferred for sorghum silage production. These hybrids produce high biomass yields, contain some grain and are typically drier at harvest than photoperiod-sensitive forage sorghum hybrids (Table 2) (Bean and McCollum, 2006). High biomass yields are accomplished through the production of long internodes and large leaves. Tall plants are often associated with lodging, so brachytic dwarf hybrids have become more popular in the industry. As mentioned previously, these hybrids have significantly smaller internode lengths, which reduce lodging. Forage hybrids also derive a significant yield from grain production, requiring proper harvesting to optimize silage digestion (Bell et al., 2015). While starch digestibility is optimized between the milk and soft-dough stages, forage moisture at harvest is also critical for optimizing silage quality. It is recommended that forages for silage be harvested at 35% dry matter. Forages harvested at higher moisture contents can result in poor fermentation and excessive effluent drainage from the silage storage structure.

3.3 Forage quality

From a livestock feed perspective, forage composition focuses on nutritive value. Analyses for livestock nutrition emphasize digestibility of the fibre and use methods that resemble an animal rumen. The overall goal is to evaluate the carbohydrates, fat, protein, minerals and vitamins available to animals for digestion and growth. Grass species are typically high in carbohydrates compared with legume forage sources. Neutral detergent fibre (NDF) is an estimation of the structural carbohydrates in forage (hemicellulose, cellulose and lignin), and acid detergent fibre (ADF) is an estimate of the cellulose and lignin. Forage analyses also include *in vitro* digestibility estimation methods such as NDF digestibility and *in vitro* dry matter digestibility (Martin and Barnes, 1980; Hatfield et al., 1994).

Attempts have been made to develop composite values that describe forage quality as a single number. Total digestible nutrients, net energy, relative feed value and relative feed quality are all composite values that have been developed to assess feed quality (Undersander et al., 1993). A range of handbooks exist that illustrate livestock response to sorghum (Brouk and Bean, 2011; Brouk, 2012).

4 Dedicated energy sorghum

Sorghum biomass yields have been evaluated throughout the United States for almost three decades (Table 3). As expected, yields vary considerably on the basis of location

Table 3 Research studies conducted to evaluate sorghum as renewable energy feedstocks

| Source | Location | Years | Dry yield | | | Description of entries[†] |
| | | | Max. | Min. | Mean | |
			Mg ha^{-1}			
Caravetta et al., 1990	Lafayette, IN	1987–88	15.9	8.3	11.8	PS
Miller and McBee, 1993	College Station, TX	1986–87	28.0	11.4	17.7	Cultivar comparison
Hallam et al., 2001	Iowa	1988–92	20.7	14.6	16.6	SWS and SS
Bean and McCollum, 2006	Bushland, TX	2000–5	10.7	7.6	9.6	Irrigated PS, multiyear means
Rooney et al., 2007	College Station, TX	1985	8.5	5.1	6.7	FS nitrogen study
Venuto and Kindiger, 2008	El Reno, OK	2004–6	33.1	23.6	29.4	PS and SS hybrids
Marsalis et al., 2009	Clovis, NM	2005–6	21.5	19.2	20.4	FS hybrid
Marsalis et al., 2010	Clovis, NM	2007–8			24.0	FS hybrid
Ottman, 2010	Maricopa, AZ	2009	15.1	7.6	12.1	FS irrigation study
Propheter et al., 2010	KS	2007–8	26.8	12.7	19.7	FS, PS and BMR
Dahlberg et al., 2011	Bushland, TX	2007	22.5	10.5	15.6	Irrigated SS and FS
Tamang et al., 2011	Lubbock, TX	2008–9	20.0	10.8	15.1	Irrigated PS N study
Maughan et al., 2012	Illinois	2009–10	39.9	13.5	24.5	ES N study, sum of two harvests
Rocateli et al., 2012	Shorter, AL	2008–9	30.1	8.0	15.0	PS and FS hybrids
Snider et al., 2012	Alabama and Arkansas	2009–10	61.0	15.0	31.8	PS hybrid

[†] FS, forage sorghum; PS, photoperiod sensitive; SWS, sweet sorghum; SS, sorghum × sudangrass; ES, energy sorghum.

and year. The lowest yields reported in the peer-reviewed literature were either rainfed or low-irrigation treatments in Texas (Bean and McCollum, 2006; Rooney et al., 2007) or Arizona (Ottman, 2010). The greatest yields with traditional forages were produced in high-yield environments with photoperiod-sensitive hybrids. The highest yield, 61 Mg ha^{-1} dry matter, was reported in Alabama in a narrow-row configuration with photoperiod-sensitive forage hybrid (Snider et al., 2012). Yields above 30 Mg ha^{-1} dry matter were reported with similar hybrids by Miller and McBee (1993), Venuto and Kindiger (2008), and Rocateli et al. (2012). The highest yields for energy sorghum were obtained in Illinois, and that sorghum outyielded traditional forage hybrids by 35% (Maughan et al., 2012).

As with digestibility in ruminant animals, lignin reduces conversion of biomass to sugars and eventually ethanol or other chemicals (Dien et al., 2001). Because lignin is highly resistant to chemical cleaving and fills in the spaces between cellulose and hemicellulose, pre-treatment processes to cleave lignin strands into more digestible segments have been evaluated in attempts to improve sugar recovery from cellulosic feedstocks (Yang and Wyman, 2004; Wyman et al., 2005; Pedersen et al., 2010; Hu et al., 2012). With the only commercial market currently being the livestock forage market, commercial-scale applications of BMR feedstocks into the biochemical markets are not widespread.

A range of in vitro lab procedures has been developed to estimate forage digestibility by ruminant animals; however, the classification of the carbohydrates as structural and non-structural differs among methods. As interest in the use of crop biomass as a source of simple sugars increased, it became apparent that different analytical methods would be needed. Although estimations of cellulose and hemicellulose could be calculated from detergent fibre methods (ADF and NDF), these methods do not often correlate with dietary or extraction methods (Wolfrum et al., 2009). Methods developed by the US DOE National Renewable Energy Laboratory (NREL) are the standards used in evaluating biomass samples for structural and non-structural carbohydrates, lignin and ash (Sluiter et al., 2005a,b, 2011).

The composition of sorghum biomass has focused largely on the structural carbohydrates: cellulose, hemicellulose and pectin. All three components are composed primarily of six sugars arranged in various complex and amorphous configurations (Heltd and Piechulla, 2011). Miller and McBee (1993) analysed forage and grain sorghum for both structural and non-structural carbohydrates (Table 4). As with mixed sorghum types, the wide range in starch values often indicates the presence of grain. Their values for hemicellulose and cellulose are similar to those from annual crops. Dien et al. (2009) reported similar data for a sample set that contained sorghum with the bmr genes, thus the reason for the low lignin values. Corredor et al. (2009) found that total structural carbohydrates in sorghum ranged from 48 to 80%, with lignin ranging from 11 to 20%. They also reported cellulose concentrations of 24–38% and hemicellulose concentrations of 18–22%. Additional work by Theerarattananoon et al. (2010) reported similar percentages of structural carbohydrates and also illustrated the range of compositions across different forage sorghum types. They reported lignin levels to be as low as 15% in a BMR forage sorghum compared with 18–20% in other forage and corn biomass samples. The BMR forage samples also had some of the highest glucan and xylan levels of the forages tested.

Dahlberg et al. (2011) conducted one of the most extensive evaluations of sorghum cultivars for yield and composition. They selected 22 cultivars that comprised both commercial forage and sweet sorghums (Table 4). The group included forage sorghum hybrids, So×Su hybrids and sweet sorghum varieties. They also included subsets of BMR, non-BMR, photoperiod-sensitive and non-photoperiod-sensitive cultivars. They found that

Table 4 Forage and energy sorghum composition from a range of studies conducted in the USA.

Source		Lignin	Cellulose	Hemicellulose	Starch	Ash	Description
		mg g^{-1}					
Miller and McBee, 1993	Max.	90.0[†]	344.0	251.0	52.0	NR[‡]	Forage sorghum and grain types
	Min.	53.0	226.0	184.0	15.0	NR	
	Mean	70.3	2663.8	204.8	29.7	NR	
Dien et al., 2009	Max.	148.8	256.0	203.0	39.0	67.6	Degrained samples, includes BMR[§] hybrids
	Min.	100.3	240.0	183.0	19.0	51.6	
	Mean	125.7	246.9	191.6	28.9	59.7	
Corredor et al., 2009	Max.	204.7	387.2	224.8	229.1	108.7	PS-BMR, PS-non-BMR, forage BMR and forage sorghum
	Min.	110.6	242.1	123.2	8.4	69.3	
	Mean	163.3	341.2	182.0	96.7	92.7	
Dahlberg et al., 2011	Max.	158.5	434.5	239.2	428.3	117.9	Forage, BMR, sorghum–sudangrass and non-grain types
	Min.	94.9	187.5	158.4	0.0	73.9	
	Mean	123.6	274.3	188.7	154.1	92.1	

[†] Lignin determination method not listed.
[‡] NR, not reported.
[§] BMR, brown midrib.

carbohydrate composition varied widely among the different cultivars. Ash ranged from 7.3 to 11.3 g g^{-1}, lignin from 9.9 to 16.3 g g^{-1}, glucans from 18.7 to 35.2 and xylans from 12.6 to 19.9 g g^{-1}.

4.1 Saccharification from sorghum biomass

Biomass to ethanol conversion, as with most plant-based biochemical production process, is driven by glucose and xylose, the two primary C6 and C5 sugars found in plant tissues. Pre-treatments and extraction rates may differ with the method and potentially by feedstock species; however, if good analytical data can be obtained for the concentrations of the C6 and C5 sugars extracted from a given amount of biomass, then the ethanol yields will be predictable. The NREL currently cites the rate for the conversion of C6 and C5 sugars to ethanol as 0.51 kg ethanol/kilogram sugar (USDOE, 2006). These conversion factors take into account the addition of water during hydrolysis. This approach has been used by numerous authors to estimate ethanol yields. Dahlberg et al. (2011) estimated that ethanol yields from forage sorghum and sorghum-sudan hybrids ranged from 336 to 440 L Mg^{-1}, with a mean of 403 L Mg^{-1}. Theerarattananoon et al. (2010) reported sorghum composition data that resulted in ethanol estimates ranging from 486 to 521 L Mg^{-1}. Estimated ethanol yields for wheat, corn, big bluestem and biomass sorghum were 473,

476, 449 and 471 L Mg^{-1}, respectively, based on composition data (Theerarattananoon et al., 2012). As previously mentioned, BMR sorghum hybrids typically have lower lignin concentrations and have been used in the livestock-feeding industry because of higher conversion rates. These characteristics carry over into cellulosic ethanol production. Dahlberg et al. (2011) reported that the highest conversion rate, 420 L Mg^{-1}, was from a BMR hybrid, but low yields across all the BMR hybrids evaluated did not result in higher ethanol yields per hectare compared with the other sorghum cultivars examined. Dien et al. (2009) reported glucose yields to be 34% higher in double mutant *bmr6* and *bmr12* hybrids compared with their near-isogenic wild types. Ethanol yields improved from 21% for *bmr6* to 43% for the double mutant compared with the wild type.

The production of organic acids, primarily acetic, propionic, butyric and lactic acids, from biomass and grain has been under parallel development in the lignocellulosic ethanol production industry but has gained less public attention (Du and Yu, 2002; Zhan et al., 2002; Carole et al., 2004; Causey et al., 2004). These acids serve as precursors to plastics, fibres and polyurethanes, which are currently produced primarily from petroleum substrates. While most of these developmental efforts have focused on a wide range of feedstocks, those with grain and sweet sorghum have met with a great deal of success because these cultivars have starches and free sugars that are easier to access compared with sugars sourced from biomass (Samuel et al., 1980; Richter and Berthold, 1998; Richter and Träger, 1994; Zhan et al., 2003; Wee et al., 2006).

4.2 Thermal conversion of sorghum biomass

The combustion of raw materials for the creation of heat has been a part of human existence for millennia. New technology in the late nineteenth century led to this energy source being harnessed to generate electricity. Coal dominated as the preferred source for heat and electricity generation until the late twentieth and early twenty-first centuries, when atmospheric CO_2 concentrations became a concern. Combustion of biomass became a topic of interest, with wood being the dominant source, and biomass from perennial and annual crops used on a smaller scale. More than 56 million megawatts of electricity are generated annually in the United States from biomass (USEIA, 2012), with wood being the dominant fuel; however, crop biomass and residues also have been used to generate electricity. The University of Minnesota Morris generates electricity with a gasification system using corn residue, wood and perennial grasses as fuel (Univ. Minnesota, Morris, 2013). Sorghum has been successfully combusted to generate electricity in California (Chromatin, 2012).

As a heat-generation fuel, sorghum has high heating values (HHVs) that are comparable to those of other renewable resources (Table 5). Sorghum biomass HHVs have been reported to range from 16.00 to 18.32 MJ kg^{-1}. One of the challenges of using annual herbaceous crop biomass as a fuel source is that they are high in alkali metals (e.g. K and Na) and halides (e.g. Cl), with amounts often significantly exceeding the 2–3% ash that is typically desired (Miles et al., 1995). High levels of these chemicals increase sintering and agglomeration in reactors and reactor beds (Ergudenler and Gahly, 1993; Miles et al., 1996; Jenkins et al., 1998; Nielsen et al., 2000; Fryda et al., 2008). The addition of kaolin, dolomite and ammonium sulphate has been reported to reduce slagging and the corrosive effects of halides such as Cl (Davidsson et al., 2002; Ohman et al., 2004; Davidsson et al., 2008; Kassman et al., 2011).

Probably the most widely known use of sorghum for electricity generation is that of sweet sorghum bagasse at ethanol plants in India, Brazil and China (Prasad et al., 2007;

Table 5 Fuel analyses for several plant biomass and fossil fuel sources from various reports

Source	Crop	Fixed carbon	Volatile material	Ash % of ash	C	H	O	N	S	High heating value	SiO in ash	K₂O in ash	Cl in ash or plant material
		% (w/w) on dry wt. basis	MJ kg⁻¹										
Fryda et al., 2008	Sweet sorghum bagasse	–	–	3.20	49.50	6.20	40.10	0.90	0.01	18.32	31.60	31.60	5.10
	Arundo donax L.	–	–	2.48	46.50	5.70	44.70	0.50	0.01	17.98	44.20	30.00	–
Hartley et al., 2011	Biomass sorghum	–	–	6.58	45.16	5.61	42.23	0.46	0.00	17.99	–	–	–
Ture et al., 1997	Sweet sorghum	–	–	2.18	44.04	6.26	–	0.22	0.08	16.83	–	–	–
	Corn stover	–	–	11.63	46.64	5.66	39.59	0.67	0.08	18.26	–	–	–
	Wheat straw	–	–	10.22	43.88	5.26	38.75	0.63	0.16	17.36	–	–	–
	Switchgrass	–	–	5.84	47.26	5.58	40.70	0.59	0.09	18.66	–	–	–
Gonzales et al., 2006	Sweet sorghum	10.30	61.30	2.70	34.00	4.50	60.23	0.80	0.02	16.00	–	–	0.45†
	Forest pellet	13.80	76.40	1.00	46.50	6.80	44.77	1.90	0.00	18.40	–	–	0.03†
	Arundo donax L.	11.40	58.40	2.20	40.30	5.30	53.13	0.40	0.07	17.40	–	–	0.80†
Channiwala and Parikh, 2002	Sugarcane bagasse	13.51	83.66	3.20	45.48	5.96	45.21	0.15	–	18.73	–	–	–
	Methane	–	–	–	74.85	25.15	–	–	–	55.35	–	–	–
	Bituminous coal	65.25	33.45	6.30	76.65	4.78	10.87	0.54	0.42	31.00	–	–	–
	Municipal solid waste	–	–	12.00	47.60	6.00	32.90	1.20	0.30	19.88	–	–	–
Jenkins and Baxter, 1998	Yard waste	13.59	66.04	20.37	41.54	4.79	31.91	0.85	0.24	16.30	59.65	2.96	0.30
	Mixed paper	7.42	84.25	8.33	47.99	6.63	36.84	0.14	0.07	20.78	28.10	0.16	0.00

† Cl reported as % of plant dry matter.

Reddy et al., 2007; Zhang et al., 2010). Cubuk et al. (2011) reported that co-combustion of sweet sorghum biomass and lignite coal resulted in similar NOx emissions but reduced C emissions. Cubuk and Heperkan (2004) noted that a similar mixture reduced pollutant concentrations in the emissions from lignite coal. Tillman (2000) pointed out that co-firing biomass with coal reduced the emission of NOx, SOx, fossil fuel CO_2 and trace minerals, such as mercury; however, co-firing biomass with coal resulted in reduced boiler efficiency in every test he reviewed.

The combustion of sweet sorghum bagasse can be quickly adopted because sugarcane bagasse has been used for centuries as a power source at sugarcane mills and also because it is readily available and eliminates the need for disposal. This system does not have the fouling and corrosion problems mentioned above because the K, Na and Cl are physically removed from the bagasse during the sugar extraction process along with the sugar. Das et al. (2004) and Jenkins et al. (1996) reported that nearly all (>92%) of the measured K, Na and Cl were removed by leaching the biomass with water.

Sorghum has been evaluated as feedstock for other thermochemical conversation processes. Pyrolysis results from sorghum bagasse and sweet sorghum biomass indicated that sorghum produced similar results as other grasses (Piskorz et al., 1998; Yin et al., 2013). Sorghum has also been studied as a feedstock for slow pyrolysis as well (Cordella et al., 2013). Yue et al. (2017) and Mafu et al. (2016) also evaluated sorghum as feedstock for torrefaction, or the production of biochar. In all cases, sorghum performed in manners similar to other annual grasses.

4.3 Anaerobic digestion of sorghum

Anaerobic digestion as a pathway to convert sorghum to biogas for heat and electricity or for transportation fuel is not widely used in the United States. Anaerobic digestion has been used for decades to reduce emissions from livestock waste systems by stimulating the biodegradable material to degrade and capturing the subsequent methane gas that is produced during degradations (Karellas et al., 2010). The inclusion of higher-carbon feedstocks that are also high in N are more beneficial than the digestion of animal manure alone. Karellas et al. (2010) listed the advantages as (i) more flexibility in siting anaerobic digesters (ADs); (ii) reduction in AD volumes as well as construction and operating costs; (iii) lower mixing rates, which reduce parasitic electricity consumption; (iv) reduced cost of feedstock logistics because of higher-density materials; and (v) production of higher-quality fertilizer in the mineral effluents.

Biogas production rates are highly dependent on the composition of the feedstock. Total solids (TS) and volatile solids (VS) are two important components to be considered when evaluating a potential feedstock (Schievano et al., 2008; Karellas et al., 2010). Schievano et al. (2008 and 2009) reported that only lab measurements of VS and oxygen demand in 20 h (OD20) were needed to predict methane production in an AD. Angelidaki et al. (1999) developed a model to estimate biogas and methane production that accounted for substrate carbohydrate and N concentrations as well as the OD20, whereas Chandra et al. (2012) indicated that a C-to-N ratio between 20 and 30 was optimal for digestion.

Sorghum biomass is high in both TS and VS (Table 6), which is optimal for the fermentation process (Richards et al., 1991). As with animal nutrition, feedstocks with lower lignin levels result in greater digestion of TS (Herrmann et al., 2011). Herrmann et al. also found that ensiling had minimal impact on AD and that most of the dry matter lost during the ensiling process was converted to lactic and acetic acid which are easily consumed in an AD. Also

Table 6 Composition and methane yields from a range of agricultural based feedstocks

		Total solids	Volatile solids	Methane
Source	Feedstock	g kg^{-1}	g kg^{-1} TS	L g VS^{-1}
Schievano et al., 2009	Cattle manure	18	799	0.058[†]
Schievano et al., 2009	Swine manure	30	602	0.126[†]
Schievano et al., 2009	Poultry litter	235	680	0.153[†]
Schievano et al., 2009	Maize silage	300	915	0.330[†]
Schievano et al., 2009	Sweet sorghum silage	200	905	0.290[†]
Richards et al., 1991	Sorghum silage	900	945	0.360
Lee et al., 2011	Corn thin stillage	73	64	0.630
Antonopoulou et al., 2008	Sweet sorghum juice	20	20	0.317[‡]
Miller and McBee, 1993	Forage sorghum	300	958	0.285
Jerger et al., 1987	Sweet sorghum (Rio)	343	945	0.400
Jerger et al., 1987	Grain sorghum (RS 610)	294	894	0.310
Jerger et al., 1987	High energy (ATx623 × Rio)	322	942	0.340

[†] Values estimated from model.
[‡] Values estimated from H_2 yields.

of interest is the use of thin stillage from grain-ethanol plants. Thin stillage is the liquid removed from wet distillers grains after ethanol production. Thin stillage is very low in dry matter (~6%) (Egg et al., 1985). Thin stillage from either corn or sorghum is quite limited in its ability to provide VS or organically available dry matter for digestion (Table 6). Sweet sorghum juice and thin stillage are both low TS and VS, suggesting that both could be used as the liquid substrate in ADs to increase methane production while reducing the water use.

Addition of plant material as AD substrates increases yields over those of manure-only substrates. Zhou et al. (2011) reported a sixfold increase in biogas production when grass silage was digested compared with cow manure alone. Forage and sweet sorghum are feedstocks that produce high yields of biogas because of their elevated content of free sugars and VS (Table 6). High biomass sorghum has been shown to be a good biogas feedstock. Claassen et al. (2004) and Antonopoulou et al. (2008) described systems in which sweet sorghum bagasse was used as a substrate by thermophilic bacteria to produce H_2 gas for biopower. High biomass yield per land area and high digestion rates result in high energy production per unit land area. Mahmood and Honermeier (2012) reported production of 8114 m^3 biogas ha^{-1} and 4333 m^3 methane ha^{-1} from sorghum. Jerger et al. (1987) reported more than 90% conversion of VS from sweet sorghum biomass to methane, with yields near 0.36 L g^{-1} VS.

5 Sweet sorghum

Sweet sorghum cultivars are forage sorghum cultivars that have been selected for their high stem juice yields and elevated sap sugar concentrations. Sweet sorghum cultivars routinely have sugar concentrations above 15% (v/v) (Wu et al., 2010). Compared with grain sorghum, which produces a large panicle of grain containing complex carbohydrates, sweet sorghum stores non-structural carbohydrates in its stems (McBee et al., 1988). Sweet sorghum was originally used in the United States for human consumption as either crystal sugar or molasses.

Sweet sorghum production in the United States has largely been concentrated in the South and occupied enough acres to warrant attempts at improvement (Walton et al., 1938). It was estimated that nearly 190 million L of sweet sorghum syrup was produced in 1920. Cultivar improvement efforts were located at USDA facilities in Meridian, MS, and Weslaco, TX, from 1950 to 1970 (Coleman, 1970; Hipp et al., 1970). As a result, sweet sorghum cultivars were bred for either molasses production or crystal sugar production (Nathan, 1978). Sugar extracted from sweet sorghum, much like that from sugarcane, is composed primarily of sucrose, glucose and fructose. Concentrations of each sugar vary among cultivars, with sucrose concentrations ranging from 65 to 93% (Corn, 2009; Kim and Day, 2011; Teetor et al., 2011; Wu et al., 2011). Glucose typically represents 75% of the remaining sugars, and fructose the balance. With increased interest in renewable fuels, research on sweet sorghum rose again in the early part of the twenty-first century, but largely on the same cultivars released from the 1950s through the 1980s (Miller and Ottman, 2010; Propheter et al., 2010; Wortmann et al., 2010; Teetor et al., 2011; Godsey et al., 2012).

Resent research indicates that sweet sorghum can attain similar fresh weights and fermentable sugars per unit of planted area as sugarcane (Table 7). The highest yields in the United States were reported in the humid, warm locations of Louisiana and South Texas and under irrigation in the desert Southwest. Yields were reduced by nearly 50% when sweet sorghum was grown under rainfed conditions in the southern and northern Great Plains, as reported by Tamang et al. (2011) and Wortmann et al. (2010). Sweet sorghum performance at locations that may be construed as the northern limit for sweet sorghum production in the United States – northeast Kansas (Propheter et al., 2010) and south central Nebraska (Wortmann et al., 2010) – indicate that water supply (stored soil water and in-season rainfall) can influence sweet sorghum yields, but these environments are also capable of producing relatively high yields.

Across all of the locations analysed, juice concentration was less affected, which could have resulted from the situation that a majority of the data reported is from M81E, an open-pollinated sweet sorghum variety developed in 1981 at the USDA facility in Meridian, MS (Broadhead et al., 1981). It is expected that sugarcane sugar extraction technology will be used to extract sugar from the sweet sorghum stalks. This assumption allows for theoretical juice and sugar yields to be calculated on the basis of moisture content and sugar concentration. In many instances, sugar concentration is measured in degrees Brix, which is a measure of dissolved solids in a solution. For our analyses, sugar yields were estimated with Eq. 1, in which sugar and juice are measured in Mg ha^{-1} (Corn, 2009)

$$\text{Sugar yield} = 0.95 \times \text{juice yield} \times 0.873 \times \frac{\text{brix}}{100} \qquad (1)$$

This relationship assumes that the expected sugar extraction rate from sweet sorghum stalks is approximately 95% (Bennett and Anex, 2009). The 0.873 is the percentage of fermentable sugars present (in degrees Brix) and is based on measurements made by Corn (2009).

Using a conversion rate for sugar to ethanol of 584 L Mg^{-1} sugar for sweet sorghum juice (Shapouri et al., 2006), 478 L Mg^{-1} biomass for sweet sorghum residual biomass (bagasse)

Table 7 Sweet sorghum biomass yield, juice yields, and juice composition from a range of studies conducted in the USA

Source	Site	Biomass	Stover[‡]	Grain[‡]	Juice[§]	Juice concentration Brix	Juice	Bagasse	Grain	Total
		Mg ha⁻¹					L ha⁻¹			
Propheter et al., 2010	KS	30.4	28.8	1.7	53.4	14.8	4390.1	13742.5	686.4	18819.0
Miller and Ottman, 2010	AZ	26.0	24.6	1.4	45.7		2472.2	11753.5	587.1	14812.7
Teetor et al., 2011	AZ	39.1	37.0	2.1	68.7	12.4	3324.5	17671.1	882.6	21878.2
Kim and Day, 2011	LA	30.0	28.4	1.6	52.7	10.5	3066.6	13561.7	677.4	17305.6
Corn, 2009	TX	43.7	40.2	3.5	74.7	13.7	3368.4	19228.3	1456.0	24052.8
Tamang et al., 2011	TX	14.9	14.1	0.8	26.2	11.3	1643.7	6735.6	336.4	8715.7
Tew et al., 2008	LA	28.6	27.1	1.6	50.3		4900.0	12942.4	646.4	18488.8
Wortmann et al., 2010	NE	9.8	9.2	0.5	17.2	9.9	1087.7	4418.8	220.7	5727.3
Wortmann et al., 2010	NE	15.9	15.1	0.9	28.0	11.2	2485.4	7208.0	360.0	10053.5

Dry yields / Ethanol[†]

[†] Ethanol yields estimated from juice yields or dry yields using the following conversion rates: juice = 584.8 L ethanol per Mg sugar; bagasse = 478 L ethanol per Mg dry matter; grain = 416 L ethanol per Mg dry matter; sugar yield ha⁻¹ = 0.95 × juice yield × 0.873 × (brix/100).

[‡] Grain and stover yields may have been estimated based on total biomass yields, subtraction and an assumed harvest index of 0.05.

[§] Theoretical juice yield estimated as the difference between fresh weights and dry weights. If not reported, moisture content was assumed to be 65%.

(Dahlberg et al., 2011), and 416 L ethanol Mg^{-1} for sweet sorghum grain (Shapouri et al., 2006) resulted in ethanol estimates ranging from 5727 to more than 24 000 L ha^{-1} (Table 7). On average, juice accounts for close to 20%, bagasse 77% and grain 3% of these ethanol values. These estimates clearly illustrate the value of large amounts of bagasse as a feedstock. The fate of bagasse will be dictated by the value of the end energy product, which can include ethanol, electricity or steam. This value proposition will also be dictated by the local infrastructure and energy needs.

Sweet sorghum hybrids and varieties are often overlooked as potential livestock feed. Their high yields of biomass and sugar make them ideal for forage products. In fact, many forage sorghum hybrids used for silage have high sugar contents but low juice yields. Adewakun et al. (1989) found that Brandes sweet sorghum resulted in similar average daily weight gain as corn silage and greater average daily weight gain compared with fescue hay. Organic matter digestibility, digestibility of crude protein and ether extracts, and gross energy digestibility were higher in steers fed sweet sorghum compared with corn silage. Caswell et al. (1983) suggested that the greater amounts of sugar and ether extracts in sweet sorghum provide more readily available energy for animal performance. Lance et al. (1964) reported similar milk yields from sweet sorghum feed compared with corn silage. Stefaniak et al. (2012) noted that sweet sorghum hybrids were higher in starch than other forage sorghum groups, suggesting that this could influence animal performance.

One advantage of sweet sorghum as a livestock feed is that the chemical producer is primarily interested in the sugar-containing juice. The by-product of this process would be the bagasse. Most studies have shown that there is very little loss of dry matter as a result of the juicing process and, in fact, that the biomass is macerated to into very fine pieces, potentially improving digestibility (Propheter et al., 2010). Being a by-product, bagasse also has the potential to reduce feed costs for the livestock producer and an additional income source for the chemical manufacturer.

6 Summary

Sorghum is one of the most genetically diverse crops used in agriculture. Forage sorghum is no exception, with the diversity increasing with the inclusion of So×Su and Su×Su hybrids. Current forage sorghum end users have the benefit of new traits, such as BMR and brachytic dwarfism, to improve digestibility and harvested yield. To date, most of these forages are used in the diets of ruminant livestock. In the past decade, an interest in using sorghum for renewable energy feedstocks has given rise to biomass sorghum, which comprises selections from forage sorghum hybrids with a focus on biomass yield and less on biomass quality. Attention to sorghum as a renewable feedstock also stimulated a renewed interest in sweet sorghum, which was developed by selecting forage sorghum hybrids with high stem-sugar content and juice yields. These hybrids have been positioned to extend the sugarcane-growing season in some environments and have also been found to produce high levels of biogas in ADs. Biomass sorghum has also been proposed as a combustion feedstock, but slagging caused by high levels of K and Cl requires additional technology to be deployed to successfully use biomass sorghum as a heat source. Forage and biomass sorghum have proven to be reliable renewable food and fibre sources, especially in marginal environments in which water is limiting.

7 Where to look for further information

Improving sorghum for any use mentioned in this chapter can be enhanced by improving the crop's adaptability to environmental stresses. Since sorghum is typically grown in marginal environments, developing more drought- and salt-tolerant sorghum inbred lines and hybrids will benefit sorghum breeding programs and end users alike. Effort by the Department of Energy through the ARPA-E TERRA program have sought to identify germplasm that is more drought and salt tolerant. This program is also focused on developing rapid phenotyping sensor arrays and platforms to decrease the time required to manually measure and evaluate such germplasm.

Since many of the characteristics that improve sorghum as a livestock feed component also improve its efficiency in renewable energy programs, improvements in yields and quality are necessary. Reducing lignin or improving digestibility while maintaining agronomic traits will result in desirable products for all sorghum based industries. Increasing seed size and starch digestibility are also important traits that will increase silage digestibility and should be a focal point of all forage sorghum breeding programs. Harvestable yield has long been a stigma in tall forage sorghum. Lodging is always the concern with these high yielding sorghum hybrids and improving standability while maintaining forage digestibility should also be a focal point in improving forage and biomass sorghum.

8 References

Adewakun, L. O., A. O. Famuyiwa, A. Felix and T. A. Omole. 1989. Growth performance, feed intake and nutrient digestibility by beef calves fed sweet sorghum silage, corn silage, and fescue hay. *J. Anim. Sci.* 67: 1341–9.

Angelidaki, I., L. Ellegaard and B. K. Ahring. 1999. A comprehensive model of anaerobic bioconversion of complex substrates to biogas. *Biotechnol. Bioeng.* 63: 363–72. doi:10.1002/(SICI)1097-0290(19990505)63:3<363::AID-BIT13>3.0.CO;2-Z.

Antonopoulou, G., H. N. Gavala, I. V. Skiadas, K. Angelopoulos and G. Lyberatos. 2008. Biofuels generation from sweet sorghum: Fermentative hydrogen production and anaerobic digestion of the remaining biomass. *Bioresour. Technol.* 99: 110–19. doi:10.1016/j.biortech.2006.11.048.

Bean, B. and T. McCollum. 2006. Summary of six years of forage sorghum variety trials. Publ. SCS-2006-04. Texas A&M AgriLife, Texas A&M Univ.: College Station.

Bell, J., Q. Xue, T. McCullom, P. Sirmon, T. Brown and D. Pietsch. 2015. 2014 Texas Panhandle Sorghum Silage Trial. Texas A&M AgriLife, Texas A&M Univ.: College Station.

Bennett, A. S. and R. P. Anex. 2009. Production, transportation and milling costs of sweet sorghum as a feedstock for centralized bioethanol production in the upper Midwest. *Bioresour. Technol.* 100: 1595–607. doi:10.1016/j.biortech.2008.09.023.

Bittinger, T. S., R. P. Cantrell and J. D. Axtell. 1981. Allelism tests of the brown-midrib mutants of sorghum. *J. Hered.* 72: 147–8.

Bout, S. and W. Vermerris. 2003. A candidate-gene approach to clone the sorghum Brown midrib gene encoding caffeic acid O-methyltransferase. *Mol. Genet. Genomics* 269: 205–14.

Broadhead, D. M., K. C. Freeman and N. Zummo. 1981. M 81E: A New Variety of Sweet Sorghum. Info. Sheet 1309. Mississippi Agric. and Forestry Exp. Stn.: Starkville.

Brouk, M. J. 2012. *Sorghum in Beef Production Feeding Guide.* United Sorghum Checkoff Program: Lubbock, TX.

Brouk, M. J. and B. Bean. 2011. *Sorghum in Dairy Cattle Production Feeding Guide*. United Sorghum Checkoff Program: Lubbock, TX.

Bucholtz, D. L., R. P. Cantrell, J. D. Axtell and V. L. Lechtenberg. 1980. Lignin biochemistry of normal and brown midrib mutant sorghum. *J. Agric. Food Chem.* 28: 1239–41. doi:10.1021/jf60232a045.

Caravetta, G. J. and K. D. Johnson. 1990. Within-row spacing influences on diverse sorghum genotypes: II. Dry matter yield and forage quality. *Agron. J.* 82: 210–15.

Carole, T. M., J. Pellegrino and M. D. Paster. 2004. Opportunities in the industrial biobased products industry. *Appl. Biochem. Biotechnol.* 115: 871–85. doi:10.1385/ABAB:115:1-3:0871.

Caswell, L. F., R. S. Kalmbacher and F. G. Martin. 1983. Yield and silage fermentation characteristics of corn, sweet sorghum and grain sorghums. *Proc. Soil Crop Sci. Soc. Fla.* 42: 139–42.

Causey, T. B., K. T. Shanmugam, L. P. Yomano and L. O. Ingram. 2004. Engineering *Escherichia* coli for efficient conversion of glucose to pyruvate. *Proc. Natl. Acad. Sci. U. S. A.* 101: 2235–40. doi:10.1073/pnas.0308171100.

Chandra, R., H. Takeuchi and T. Hasegawa. 2012. Methane production from lignocellulosic agricultural crop wastes: A review in context to second generation of biofuel production. *Renew. Sustain. Energy Rev.* 16: 1462–76. doi:10.1016/j.rser.2011.11.035.

Childs K. L., F. R. Miller, M. M. Cordonnier-Pratt, L. H. Pratt, P. W. Morgan and J. E. Mullet. 1997. Plant Physiol. 113: 611–19. doi:http://dx.doi.org/10.1104/pp.113.2.611.

Chromatin. 2012. Constellation energy and chromatin announce partnership to test sorghum biomass as fuel to generate power. Press release, Chromatin, Inc., 21 September 2011. http://www.chromatininc.com/news/Constellation-Energy-and-Chromatin-Announce-Partnership-to-Test-Sorghum-Biomass-as-Fuel-to-Generate-Power (Accessed 9 May 2013).

Claassen, P. A. M., T. de Vrije and M. A. W. Budde. 2004. Biological hydrogen production from sweet sorghum by thermophilic bacteria. *Proceedings of the 2nd World Conference on Biomass for Energy, Industry, and Climate Protection, Rome Italy*. 10–14 May. ETA-Renewable Energies, Florence, Italy, and WIP, Munich, Germany, pp. 1522–5.

Coleman, O. H. 1970. Syrup and sugar from sweet sorghum. In J. S. Wall and W. M. Ross (Eds), *Sorghum Production and Utilization*. AVI Publishing: Westport, CT, pp. 416–41.

Cook, O. F. 1915. Brachysm, a hereditary deformity of cotton and other plants. *J. Agric.Res.* 3: 387–99.

Cordella, M., C. Berrueco, F. Santarelli, N. Paterson, R. Kandiyoti and M. Millan, 2013. Yields and ageing of the liquids obtained by slow pyrolysis of sorghum, switchgrass and corn stalks. *J. Anal. Appl. Pyrolysis*, 104: 316–24. ISSN: 0165-2370, http://dx.doi.org/10.1016/j.jaap.2013.07.001.

Corn, R. J. 2009. Heterosis and composition of sweet sorghum. Ph.D. diss. Texas A&M Univ.: College Station. http://hdl.handle.net/1969.1/ETD-TAMU-2009-12-7409.

Corredor, D. Y., J. M. Salazar, K. L. Hohn, S. Bean, B. Bean and D. Wang. 2009. Evaluation and characterization of forage sorghum as feedstock for fermentable sugar production. *Appl. Biochem. Biotechnol.* 158: 164–79. doi:10.1007/s12010-008-8340-y.

Craufurd, P. Q., V. Mahalakshmi, F. R. Bidinger, S. Z. Mukuru, J. Chantereau, P. A. Omanga, A. Qi, E. H. Roberts, R. H. Ellis, R. J. Summerfield and G. L. Hammer. 1999. Adaptation of sorghum: Characterisation of genotypic flowering responses to temperature and photoperiod. *Theor. Appl. Genet.* 99: 900–11. doi:10.1007/s001220051311.

Cubuk, M. H. and H. A. Heperkan. 2004. Investigation of pollutant formation of sweet sorghum–lignite (Orhaneli) mixtures in fluidised beds. *Biomass Bioenergy* 27: 277–87. doi:10.1016/j.biombioe.2004.02.001.

Cubuk, M. H., D. B. Ozkan and O. Emanet. 2011. NOx formation of co-combustion of sweet sorghum–ignite (Orhaneli) mixtures in fluidized beds. In H. Gökçekus, U. Türker and J. W. LaMoreaux (Eds), *Survival and Sustainability*. Springer: Berlin, pp. 931–41.

Dahlberg, J., E. Wolfrum, B. W. Bean and W. L. Rooney. 2011. Compositional and agronomic evaluation of sorghum biomass as a potential feedstock for renewable fuels. *J. Biobased Mat. Bioenergy* 5: 507–13. doi:10.1166/jbmb.2011.1171.

Das, P., A. Ganesh and P. Wanikar. 2004. Influence of pretreatment for deashing of sugarcane bagasse on pyrolysis products. *Biomass Bioenergy* 27: 445–57.

Davidsson, K. O., L. E. Åmand, B. M. Steenari, A. L. Elled, D. Eskilsson and B. Leckner. 2008. Countermeasures against alkali-related problems during combustion of biomass in a circulating fluidized bed boiler. *Chem. Eng. Sci.* 63: 5314–29. doi:10.1016/j.ces.2008.07.012.

Davidsson, K. O., J. G. Korsgren, J. B. B. Perrersson and U. Jaglid. 2002. The effects of fuel washing techniques on alkali release from biomass. *Fuel* 81: 137–42. doi:10.1016/S0016-2361(01)00132-6.

Dien, B. S., G. Sarath, J. F. Pedersen, S. E. Sattler, H. Chen, D. L. Funnell-Harris, N. N. Nichols and M. A. Cotta. 2009. Improved sugar conversion and ethanol yield for forage sorghum (*Sorghum bicolor* L. Moench) lines with reduced lignin contents. *BioEnergy Res.* 2: 153–64. doi:10.1007/s12155-009-9041-2.

Dien, B. S., N. N. Nichols and R. J. Bothast. 2001. Recombinant *Escherichia coli* engineered for production of L-lactic acid from hexose and pentose sugars. *J. Ind. Microbiol. Biotechnol.* 27: 259–64. doi:10.1038/sj.jim.7000195.

Dingkuhn, M., M. Kouressy, M. Vaksmann, B. Blerget and J. Chantereau. 2008. A model of sorghum photoperiodism using the concept of threshold-lowering during prolonged appetence. *Eur. J. Agron.* 28: 74–89. doi:10.1016/j.eja.2007.05.005.

Doi K, T. Izawa and T. Fuse. 2004. Ehd1, a B-type response regulator in rice, confers short-day promotion of flowering and controls FT-like gene expression independently of Hd1. *Genes Dev.* 18(8): 926–36. doi:10.1101/gad.1189604.

Du, G. and J. Yu. 2002. Green technology for conversion of food scraps to biodegradable thermoplastic polyhydroxyalkanoates. *Environ. Sci. Technol.* 36: 5511–16. doi:10.1021/es011110o.

Egg, R. P., J. M. Sweeten, and C. G. Coble. 1985. Grain sorghum stillage recycling: Effect on ethanol yield and stillage quality. *Biotechnol. Bioeng.* 27: 1735–8.

Ergudenler, A. and A. E. Gahly. 1993. Agglomeration of silica sand in a fluidized bed gasifier operating on straw. *Biomass Bioenergy* 4: 135–47. doi:10.1016/0961-9534(93)90034-2.

Fritz, J. O., K. J. Moore and E. H. Jaster. 1990. Digestion kinetics and cell wall composition of brown midrib sorghum × sudangrass morphological components. *Crop Sci.* 30: 213–19. doi:10.2135/cropsci1990.0011183X003000010046x.

Fryda, L. E., K. D. Panopoulos and E. Kakaras. 2008. Agglomeration in fluidised bed gasification of biomass. *Powder Technol.* 181: 307–20.

Godsey, C. B., J. Linneman, D. Bellmer and R. Hunke. 2012. Developing row spacing and plant density recommendations for rainfed sweet sorghum production in the Southern Plains. *Agron. J.* 104: 280–6. doi:10.2134/agronj2011.0289.

Goud, J. V. and M. Vasudev Rao. 1977. Inheritance of height in sorghum. *Genet. Agrar.* 31: 39–51.

Hallam, A., I. C. Anderson and D. R. Buxton. 2001. Comparative economic analysis of perennial, annual, and intercrops for biomass production. *Biomass Bioenergy* 21: 407–24.

Hanna, W. W., W. G. Monson and T. P. Gaines. 1981. In vitro dry matter digestibility, total sugars, and lignin measurements on normal and brown midrib (*bmr*) sorghums at various stages of development. *Agron. J.* 73: 1050–2. doi:10.2134/agronj1981.00021962007300060034x.

Hart, G., K. Schertz, Y. Peng and N. Syed. 2001. Genetic mapping of *Sorghum bicolor* (L.) Moench QTLs that control variation in tillering and other morphological characters. *Theor. Appl. Genet.* 103: 1232–42. doi:10.1007/s001220100582.

Hatfield, R. D., J. G. Jung, J. Ralph, D. R. Buxton and P. J. Weimer. 1994. A comparison of the insoluble residues produced by the Klason lignin and acid detergent lignin procedures. *J. Sci. Food Agric.* 65: 51–8. doi:10.1002/jsfa.2740650109.

Hartley, B. E., J. D. Gibson, J. A. Thomasson and S. W. Searcy. 2011. Moisture loss and ash characterization of high-tonnage sorghum. Paper presented at ASABE Annual International Meeting, Louisville, KY. 7–10 August. Paper 1111528.

Heldt, H. W. and B. Piechulla. 2011. *Plant Biochemistry*. 4th ed. Academic Press: Burlington, MA.

Herrmann, C., M. Heiermann and C. Idler. 2011. Effects of ensiling, silage additives and storage period on methane formation of biogas crops. *Bioresour. Technol.* 102: 5153–61. doi:10.1016/j.biortech.2011.01.012.

Hill, J. D. 1976. Predicting the natural drying of hay. *Agric. Meteorol.* 17: 195–204. doi:10.1016/0002-1571(76)90055-8.

Hilley J., S. Truong, S. Olson, D. Morishige and J. Mullet. 2016. Identification of Dw1, a regulator of sorghum stem internode length. *PLoS One* 11(3): e0151271. doi:10.1371/journal.pone.0151271.

Hipp, B. W., W. R. Cowley, C. J. Gerard and B. A. Smith. 1970. Influence of solar radiation and date of planting on yield of sweet sorghum. *Crop Sci.* 10: 91–2. doi:10.2135/cropsci1970.0011183X001000010033x.

Howell, T. A, S. R. Evett, J. A. Tolk, K. S. Copeland, P. D. Colaizzi and P. H. Gowda. 2008. Evapotranspiration of corn and forage sorghum for silage. *Wetting Front.* 10(1): 3–11

Hu, G., C. Cateto, Y. Pu, R. Samuel and A. J. Ragauskas. 2012. Structural characterization of switchgrass lignin after ethanol organosolv pretreatment. *Energy Fuels* 26: 740–5. doi:10.1021/ef201477p.

Hussein, M. M. and A. K. Alva. 2014. Growth, yield, and water use efficiency of forage sorghum as affected by N, P, K fertilizer and deficit irrigation. *Am. J. Plant Sci.* 5: 2134–40. doi:10.4236/ajps.2014.513225.

Jenkins, B. M., R. R. Bakker and J. B. Wei. 1996. On the properties of washed straw. *Biomass Bioenergy* 10: 177–200. doi:10.1016/0961-9534(95)00058-5.

Jenkins, B. M., L. L. Baxter, T. R. Miles Jr. and T. R. Miles. 1998. Combustion properties of biomass. *Fuel Process. Technol.* 54: 17–46. doi:10.1016/S0378-3820(97)00059-3.

Jerger, D. E., D. P. Chynoweth and H. R. Isaacson. 1987. Anaerobic digestion of sorghum biomass. *Biomass* 14: 99–113.

Karellas, S., I. Boukis and G. Kontopoulos. 2010. Development of an investment decision tool for biogas production from agricultural waste. *Renew. Sustain. Energy Rev.* 14: 1273–82. doi:10.1016/j.rser.2009.12.002.

Kassman, H., M. Brostrom, M. Berg and L. Amand. 2011. Measures to reduce chlorine in deposits: Application in a large-scale circulating fluidised bed boiler firing biomass. *Fuel* 90: 1325–34. doi:10.1016/j.fuel.2010.12.005.

Kemanian, A. R., C. O. Stockle, D. R. Huggins and L. M. Viega. 2007. A simple method to estimate harvest index in grain crops. *Field Crops Res.* 103: 208–16. doi:10.1016/j.fcr.2007.06.007.

Kim, M. and D. F. Day. 2011. Composition of sugar cane, energy cane, and sweet sorghum suitable for ethanol production at Louisiana sugar mills. *J. Ind. Microbiol. Biotechnol.* 38: 803–7. doi:10.1007/s10295-010-0812-8.

Klein, R. R., F. R. Miller, D. V. Dugas, P.J. Brown, M. Burrell, P.E. Klein. 2015. Allelic variants in the PRR37 gene and the human-mediated dispersal and diversification of sorghum. *Theor. Appl. Genet.* 128: 1669. doi:10.1007/s00122-015-2523-z.

Klein R. R., J. E. Mullet, D. R. Jordan, F. R. Miller, W. L. Rooney and M. A. Menz. 2008. The effect of tropical sorghum conversion and inbred development on genome diversity as revealed by high-resolution genotyping. *Crop Sci.* 48: 12–26. doi:10.2135/cropsci2007.06.0319tpg.

Lance, R. D., D. C. Foss, C. R. Kruegar, B. R. Baumgardt and R. R. Niedermeir. 1964. Evaluation of corn and sorghum silages on the basis of milk production and digestibility. *J. Dairy Sci.* 47: 254–7. doi:10.3168/jds.S0022-0302(64)88635-5.

Lee, P, J. Bae, J. Kim and W. Chen. 2011. Mesophilic anaerobic digestion of corn thin stillage: A technical and energetic assessment of the corn-to-ethanol industry integrated with anaerobic digestion. *J. Chem. Technol. Biotechnol.* 86: 1514–20.

Li X., X. Li, E. Fridman, T. T. Tesso and J. Yu. 2015. Dissecting repulsion linkage in the dwarfing gene Dw3 region for sorghum plant height provides insights into heterosis. *Proc. Natl. Acad. Sci. U. S. A.* 112: 11823–8. doi:10.1073/pnas.1509229112.

Mafu, L. D., H. W. J. P. Neomagus, R. C. Everson, M. Carrier, C. A. Strydom and J. R. Bunt. 2016. Structural and chemical modifications of typical South African biomasses during torrefaction. *Bioresour. Technol.* 202: 192–7. ISSN: 0960-8524, http://dx.doi.org/10.1016/j.biortech.2015.12.007 (http://www.sciencedirect.com/science/article/pii/S0960852415016375).

Mahmood, A. and B. Honermeier. 2012. Chemical composition and methane yield of sorghum cultivars with contrasting row spacing. *Field Crops Res.* 128: 27–33. doi:10.1016/j.fcr.2011.12.010.

Marsalis, M. A., S. V. Angadi and F. E. Contreras-Govea. 2010. Dry matter yield and nutritive value of corn, forage sorghum, and BMR forage sorghum at different plant populations and nitrogen rates. *Field Crops Res.* 116: 52–7. doi:10.1016/j.fcr.2009.11.009.

Marsalis, M. A., S. V. Angadi, F. E. Contreras-Govea and R. E. Kirksey. 2009. Harvest Timing and Byproduct Addition Effects on Corn and Forage Sorghum Silage Grown Under Water Stress. Bull. 799. New Mexico State Univ.: Las Cruces.

Martin, G. C. and R. F. Barnes. 1980. Prediction of energy digestibility of forages with in vitro rumen fermentation and fungal enzyme systems. In W. G. Pigden, C. C. Balch and M. Graham (Eds), *Standardization of Analytical Methodology for Feeds.* Publ. IDR-134e. Intl. Res. Dev. Cent.: Ottawa, ON.

Maughan, M., T. Voigt, A. Parrish, G. Bollero, W. Rooney and D. K. Lee. 2012. Forage and energy sorghum response to nitrogen fertilization in central and southern Illinois. *Agron. J.* 104: 1032–40. doi:10.2134/agronj2011.0408.

McBee, G. G., R. A. Creelman and F. R. Miller. 1988. Ethanol yield and energy potential of stems from a spectrum of sorghum biomass types. *Biomass* 17: 203–11. doi:10.1016/0144-4565(88)90114-X.

McCuistion, K. C., B. W. Bean and F. T. McCollum. 2009. Yield and water-use efficiency response to irrigation level of brown midrib, non-brown midrib, and photoperiod-sensitive forage sorghum cultivars. *Forage Grazinglands* 7. doi:10.1094/FG-2009-0909-01-RS.

McCuistion, K. C., B. W. Bean and F. T. McCollum. 2010. Nutritional composition response to yield differences in brown midrib, non-brown midrib, and photoperiod sensitive forage sorghum cultivars. *Forage Grazinglands* 8. doi:10.1094/FG-2010-0428-01-RS

Miles, T. R., T. R. Miles Jr., L. L. Baxter, R. W. Bryers, B. M. Jenkins and L. L. Oden. 1995. Alkali Deposits Found in Biomass Power Plants. A Preliminary Investigation of Their Extent and Nature. Publ. NREL/TP-433–8142. Vol. 2. Sandia Natl. Lab.: Albuquerque, NM.

Miles, T. R., T. R. Miles Jr., L. L. Baxter, R. W. Bryers, B. M. Jenkins and L. L. Oden. 1996. Boiler deposits from firing biomass fuels. *Biomass Bioenergy* 10: 125–38. doi:10.1016/0961-9534(95)00067-4.

Miller, A. N. and M. J. Ottman. 2010. Irrigation frequency effects on growth and ethanol yield in sweet sorghum. *Agron. J.* 102: 60–70. doi:10.2134/agronj2009.0191.

Miller, F. R. and G. G. McBee. 1993. Genetics and management of physiological systems of sorghum for biomass production. *Biomass Bioenergy* 5: 41–9. doi:10.1016/0961-9534(93)90006-P.

Mullet, J. E., W. L. Rooney, P. E. Klein, D. Morishige, R. Murphy and J. A. Brady. 2010. Discovery and utilization of sorghum genes (*ma5/ma6*). US Patent application 20100024065 A1. Published: 28 January.

Mullet J. E. and W. L. Rooney. 2013. Method for production of sorghum hybrids with selected flowering times. US Patent 20130298274 A1.

Multani, D. S., S. P. Briggs, M. A. Chamberlin, J. J. Blakeslee, A. S. Murphy and G. S. Johal. 2003. Loss of an MDR Transporter in Compact Stalks of Maize br2 and Sorghum dw3 Mutants. *Science* 302(5642): 81–84. doi:10.1126/science.1086072.

Murphy R. L., R. R. Klein, D. T. Morishige, J. A. Brady, W. L. Rooney, F. R. Miller, D. V. Dugas, P. E. Klein and J. E. Mullet. Coincident light and clock regulation of pseudoresponse regulator protein 37 (PRR37) controls photoperiodic flowering in sorghum. *PNAS* 108(39): 16469–74. published ahead of print 19 September 2011, doi:10.1073/pnas.1106212108.

Murphy, R. L., D. T. Morishige, J. A. Brady, W. L. Rooney, S. Yang, P. E. Klein and J. E. Mullet. 2014. Ghd7 (Ma6) represses sorghum flowering in long days: Ghd7 alleles enhance biomass accumulation and grain production. *Plant Genome* 7. doi:10.3835/plantgenome2013.11.0040; doi:10.3835/plantgenome2013.11.0040.

Narayanan, S., R. M. Aiken, P. V. V. Prasad, Z. Xin and J. Yu. 2013. Water and radiation use efficiencies in sorghum. *Crop Sci.* 105:649–56.

Nathan, R. A. (Ed.). 1978. Fuels from Sugar Crops: Systems Study for Sugarcane, Sweet Sorghum, and Sugar Beets. Tech. Info. Cent. US DOE 71D-22781. US DOE: Washington, DC.

Nielsen, H. P., F. J. Frandsen, K. Dam-Johansen and L. L. Baxter. 2000. The implications of chlorine-associated corrosion on the operation of biomass-fired boilers. *Prog. Energy Combust. Sci.* 26:283–98. doi:10.1016/S0360-1285(00)00003-4.

Ohman, M., D. Bostrom, A. Nordin and H. Hedman. 2004. Effect of kaolin and limestone addition on slag formation during combustion of wood fuels. *Energy Fuels* 18: 1370–6. doi:10.1021/ef040025+.

Ottman, M. J. 2010. Water use efficiency of forage sorghum grown with sub-optimal irrigation. In *Forage and Grain Report*. Univ. of Ariz. Coop. Ext. Publ. AZ1526. Univ. of Arizona: Tucson, pp. 26–9. http://cals.arizona.edu/pubs/crops/az1526/az1526d.pdf.

Pedersen, J. F., D. L. Funnell, J. J. Toy, A. L. Oliver and R. J. Grant. 2006a. Registration of -Atlas *bmr12*' forage sorghum. *Crop Sci.* 46: 478.

Pedersen, J. F., D. L. Funnell, J. J. Toy, A. L. Oliver and R. J. Grant. 2006b. Registration of seven forage sorghum genetic stocks near-isogenic for the brown midrib genes *bmr6* and *bmr12*. *Crop Sci.* 46: 490–1.

Pedersen, J. F., D. L. Funnell, J. J. Toy, A. L. Oliver and R. J. Grant. 2006c. Registration of 12 grain sorghum genetic stocks near-isogenic for the brown midrib genes *bmr6* and *bmr12*. *Crop Sci.* 46: 491–2.

Pedersen, J. F., J. J. Toy, D. L. Funnell, S. E Sattler, A. L. Oliver and R. A. Grant. 2008. Registration of BN611, A/BN612, and RN613 sorghum genetic stocks with stacked *bmr6* and *bmr12* genes. *J. Plant Reg.* 2: 258–62.

Pedersen, M., A. Vikso-Nielsen and A. S. Meyer. 2010. Monosaccharide yields and lignin removal from wheat straw in response to catalyst type and pH during mild thermal pretreatment. *Process Biochem.* 45: 1181–6. doi:10.1016/j.procbio.2010.03.020.

Piskorz J., P. Majerski, D. Radlein, D. S. Scott and A. V. Bridgwater, 1998. Fast pyrolysis of sweet sorghum and sweet sorghum bagasse. *J. Anal. Appl. Pyrolysis* 46:15–29. ISSN: 0165-2370, http://dx.doi.org/10.1016/S0165-2370(98)00067-9.

Porter, K. S., J. D. Axtell, V. L. Lechtenberg and V. F. Colenbrander. 1978. Phenotype, fiber composition, and in vitro dry matter disappearance of chemically induced brown midrib (*bmr*) mutants of sorghum. *Crop Sci.* 18: 205–8. doi:10.2135/cropsci1978.0011183X001800020002x.

Prasad, S., A. Singh, N. Jain and H. C. Joshi. 2007. Ethanol production from sweet sorghum syrup for utilization as automotive fuel in India. *Energy Fuels* 21: 2415–20. doi:10.1021/ef060328z.

Propheter, J. L., S. A. Staggenborg, X. Wu and D. Wang. 2010. Performance of annual and perennial biofuel crops: Yield during the first two years. *Agron. J.* 102: 806–14. doi:10.2134/agronj2009.0301.

Quinby, J. R. 1974. The genetic control of flowing and growth in sorghum. *Adv. Agron.* 25: 125–62. doi:10.1016/S0065-2113(08)60780-4.

Quinby J. R. and R. E. Karper. 1961. Inheritance of duration of growth in the Milo Group of Sorghum1. *Crop Sci.* 1: 8–10. doi:10.2135/cropsci1961.0011183X000100010004x.

Quinby, J. R. and R. E. Karper. 1954. Inheritance of height in sorghum. *Agron. J.* 46: 211–16. doi:10.2134/agronj1954.00021962004600050007x.

Rajan, N. and S. Maas. 2007. Comparative evaluation of actual crop water use of forage sorghum and corn for silage. Report as of FY2007 for 2007TX271B. Dep. of the Interior, USGS. http://water.usgs.gov/wrri/07grants/2007TX271B.html (accessed 1 June 2015).

Reddy, B. V. S., A. Ashok Kumar and S. Ramesh. 2007. Sweet sorghum: A water saving bio-energy crop. Paper presented at the International conference on Linkages between Energy and Water Management for Agriculture in Developing Countries, ICRISAT campus, Hyderabad, India. 29–30 January.

Richards, B. K., R. J. Cummings, W. J. Jewell and F. G. Herndon. 1991. High solid anaerobic methane fermentation of sorghum and cellulose. *Biomass Bioenergy* 1: 47–53. doi:10.1016/0961-9534(91)90051-D.

Richter, K. and C. Berthold. 1998. Biotechnological conversion of sugar and starchy crops into lactic acid. *J. Agric. Eng. Res.* 71: 181–91. doi:10.1006/jaer.1998.0314.

Richter, K. and A. Träger. 1994. L(+)-lactic acid from sweet sorghum by submerged and solid-state fermentations. *Acta Biotechnol.* 14: 367–78. doi:10.1002/abio.370140409.

Rocateli, A. C., R. L. Raper, K. S. Balkcom, F. J. Arriaga and D. I. Bransby. 2012. Biomass sorghum production and components under different irrigation/tillage systems for the southeastern U.S. Ind. *Crops Prods.* 36:589–98. doi:10.1016/j.indcrop.2011.11.007.

Rooney, W. L. and S. Aydin. 1999. Genetic control of a photoperiod-sensitive response in sorghum bicolor (L.) moench. *Crop Sci.* 39: 397–400. doi:10.2135/cropsci1999.0011183X0039000200016x.

Rooney, W. L., J. Blumenthall, B. Bean and J. E. Mullet. 2007. Designing sorghum as a dedicated bioenergy feedstock. *Biofuels Bioprod. Biorefin.* 1: 147–57. doi:10.1002/bbb.15.

Saballos, A., G. Ejeta, E. Sanchez, C. Kang and W. Vermerris. 2009. A genomewide analysis of the cinnamyl alcohol dehydrogenase family in sorghum [*Sorghum bicolor* (L.) Moench] identifies SbCAD2 as the Brown midrib6 gene. *Genetics* 181: 783–95.

Saballos, A., S. E. Sattler, E. Sanchez, T. P. Foster, Z. Xin and et al. 2012. Brown midrib2 (*bmr2*) encodes the major 4-coumarate: Coenzyme A ligase involved in lignin biosynthesis in sorghum (*Sorghum bicolor* (L.) Moench). *Plant J.* 70: 818–30.

Saballos, A., W. Vermerris, L. Rivera and G. Ejeta, 2008 Allelic association, chemical characterization and saccharification properties of brown midrib mutants of sorghum (*Sorghum bicolor* (L.) Moench). *Bioenerg. Res.* 1:193–204. doi:10.1007/s12155-008-9025-7.

Saeed, I. A. M. and A. H. El-Nadi. 1998. Forage sorghum yield and water use efficiency under variable irrigation. *Irrig. Sci.* 18: 67–71. doi:10.1007/s002710050046.

Samuel, W. A., Y. Y. Lee and W. B. Anthony. 1980. Lactic acid fermentation of crude sorghum extract. *Biotechnol. Bioeng.* 22: 757–77. doi:10.1002/bit.260220404.

Sattler, S. E., A. J. Saathoff, E. J. Haas, N. A. Palmer, D. L. Funnell-Harris and et al. 2009. A nonsense mutation in a cinnamyl alcohol dehydrogenase gene is responsible for the sorghum brown midrib6 phenotype. *Plant Physiol.* 150: 584–95.

Schievano, A., M. Pognani, G. D'Imporzano and F. Adani. 2008. Predicting anaerobic biogasification potential of ingestates and digestates of a full-scale biogas plant using chemical and biological parameters. *Bioresour. Technol.* 99: 8112–17. doi:10.1016/j.biortech.2008.03.030.

Schievano, A., B. Scaglia, G. D'Imporzano, L. Malagutti, A. Gozzi and F. Adani. 2009. Prediction of biogas potentials using quick laboratory analyses: Upgrading previous models for application to heterogeneous organic matrices. *Bioresour. Technol.* 100:5777–82. doi:10.1016/j.biortech.2009.05.075.

Shapouri, H., M. Salassi and J. Nelson. 2006. *The Economic Feasibility of Ethanol Production from Sugar in the United States.* USDA: Washington, DC. http://www.usda.gov/oce/reports/energy/EthanolSugarFeasibilityReport3.pdf (Accessed 6 March 2016).

Sluiter, A., B. Hames, R. Ruiz, C. Scarlata, J. Sluiter, and D. Templeton. 2005a. Determination of Ash in Biomass, Laboratory Analytical Procedure. Tech. Rep. 510-42622. National Renewable Energy Lab: Golden, CO.

Sluiter, A., B. Hames, R. Ruiz, C. Scarlata, J. Sluiter, D. Templeton, and D. Crocker. 2011. Determination of Structural Carbohydrates and Lignin in Biomass, Laboratory Analytical Procedure. Tech. Rep. 510-42618. National Renewable Energy Lab: Golden, CO.

Sluiter, A., R. Ruiz, C. Scarlata, J. Sluiter and D. Templeton. 2005b. Determination of Extractives in Biomass, Laboratory Analytical Procedure. Tech. Rep. 510-42619. National Renewable Energy Lab: Golden, CO.

Snider, J. L., R. L. Raper and E. B. Schwab. 2012. The effect of row spacing and seeding rate on biomass production and plant stand characteristics on non-irrigated photoperiod-sensitive sorghum (*Sorghum bicolor* (L.) Moench). *Ind. Crops Prod.* 37: 527–535. doi:10.1016/j.indcrop.2011.07.032.

Staggenborg, S. A., K. C. Dhuyvetter and W. B. Gordon. 2008. Grain sorghum and corn comparisons: Yield, economic, and environmental responses. *Agron. J.* 100: 1600–4. doi:10.2134/agronj2008.0129.

Stefaniak, T. R., J. A. Dahlberg, B. W. Bean, N. Dighe, E. J. Wolfrum and W. L. Rooney. 2012. Variation in biomass composition components among forage, biomass, sorghum-sudangrass, and sweet sorghum types. *Crop Sci.* 52: 1949–54. doi:10.2135/cropsci2011.10.0534.

Tamang, P. L., K. F. Bronson, A. Malapati, R. Schwartz, J. Johnson and J. Moore-Kucera. 2011. Nitrogen requirement for ethanol production from sweet sorghum and photoperiod sensitive sorghums in the Southern High Plains. *Agron. J.* 103: 431–40. doi:10.2134/agronj2010.0288.

Teetor, V. H., D. V. Duclos, E. T. Wittenberg, K. M. Young, J. Chawhuaymak, M. R. Riley and D. T. Ray. 2011. Effects of planting date on sugar and ethanol yield of sweet sorghum grown in Arizona. *Ind. Crops Prod.* 34: 1293–300. doi:10.1016/j.indcrop.2010.09.010

Tew, T. L., R. M Cobill and E. P. Richard. 2008. Evaluation of sweet sorghum and sorghum × sudangrass hybrids as feedstocks for ethanol production. *Bioenerg. Res.* 1: 147–52.

Theerarattananoon, K., X. Wu, S. Staggenborg, R. Propheter, R. Madl and D. Wang. 2010. Evaluation and characterization of sorghum biomass as feedstock for sugar production. *Trans. ASABE* 53: 509–25. doi:10.13031/2013.29561.

Theerarattananoon, K., F. Xu, J. Wilson, S. Staggenborg, L. McKinney, P. Vadlani, Z. Pei and D. Wang. 2012. Effect of pelleting conditions on chemical composition and sugar yield of corn stover, big bluestem, wheat straw and sorghum stalk pellets. *Bioprocess Biosyst. Eng.* 35: 615–23. doi:10.1007/s00449-011-0642-8.

Tillman, D. A. 2000. Biomass cofiring: The technology, the experience, the combustion consequences. *Biomass Bioenergy* 19: 365–84. doi:10.1016/S0961-9534(00)00049-0.

Undersander, D., D. R. Mertens and N. Theix. 1993. *Forage Analyses Procedures.* Natl. Forage Testing Assoc.: Omaha, NE.

University of Minnesota, Morris. 2013. *Renewable Energy.* University of Minnesota: Morris. https://www.morris.umn.edu/sustainability/renewable/(Accessed 1 April 2016).

US Department of Energy (USDOE). 2006. Theoretical Ethanol Yield Calculator. USDOE: Washington, DC. http://www1.eere.energy.gov/biomass/ethanol_yield_calculator.html (accessed 30 April 2013).

US Energy Information Administration (USEIA). 2012. *Electric Power Monthly.* USDOE: Washington, DC, http://www.eia.gov/electricity/annual/archive/03482012.pdf (Accessed 30 March 2016).

Venuto, B. and B. Kindiger. 2008. Forage and biomass feedstock production from hybrid forage sorghum and sorghum-sudangrass hybrids. *Grassl. Sci.* 54: 189–96. doi:10.1111/j.1744-697X.2008.00123.x.

Walton, C. F. Jr., E. K. Ventre and S. Byall. 1938. Farm Production of Sorgo Sirup. Farmers' Bull. 1791. USDA: Washington, DC.

Wee, J. Y., J. N. Kim and H. W. Ryu. 2006. Biotechnological production of lactic acid and its recent applications. *Food Technol. Biotechnol.* 44(2): 163–172.

Wolfrum, E. J., A. J. Lorenz, and N. deLeon. 2009. Correlating detergent fiber analyses and dietary fiber analysis data for corn stover collected by NIRS. *Cellulose* 16(4): 577–85. DOI:10.1007/s10570-009-9318-9.

Wortmann, C. S., A. J. Liska, R. B. Ferguson, D. J. Lyon, R. N. Klein and I. Dweikat. 2010. Dryland performance of sweet sorghum and grain crops for biofuel in Nebraska. *Agron. J.* 102: 319–26. doi:10.2134/agronj2009.0271.

Wu, L., M. Arakane, M. Ike, M. Wada, T. Takai, M. Gau and K. Tokuyasu. 2011. Low temperature alkali pretreatment for improving enzymatic digestibility of sweet sorghum bagasse for ethanol production. *Bioresour. Technol.* 102:4793–9. doi:10.1016/j.biortech.2011.01.023.

Wu, X. R., S. A. Staggenborg, J. L. Propheter, W. L. Rooney, J. Yu and D. Wang. 2010. Features of sweet sorghum juice and their performance in ethanol fermentation. *Ind. Crops Prod.* 31: 164–70. doi:10.1016/j.indcrop.2009.10.006.

Wyman, C. E., B. E. Dale, R. T. Elander, M. Holtzapple, M. R. Ladisch and Y. Y. Lee. 2005. Comparative sugar recovery data from laboratory scale application of leading pretreatment technologies to corn stover. *Bioresour. Technol.* 96: 2026–32. doi:10.1016/j.biortech.2005.01.018.

Yang, B. and C. E. Wyman. 2004. Effect of xylan and lignin removal by batch and flowthrough pretreatment on the enzymatic digestibility of corn stover cellulose. *Biotechnol. Bioeng.* 86(1): 88–98. doi:10.1002/bit.20043.

Yang S, R. L. Murphy, D. T. Morishige, P. E. Klein and W. L. Rooney. 2014. Sorghum Phytochrome B Inhibits Flowering in Long Days by Activating Expression of SbPRR37 and SbGHD7, Repressors of SbEHD1, SbCN8 and SbCN12. *PLoS One* 9(8): e105352. doi:10.1371/journal.pone.0105352.

Yin R., R. Liu, Y. Mei, W. Fei and X. Sun, 2013. Characterization of bio-oil and bio-char obtained from sweet sorghum bagasse fast pyrolysis with fractional condensers. *Fuel* 112: 96–104. ISSN: 0016-2361, http://dx.doi.org/10.1016/j.fuel.2013.04.090.

Yue, Y., H. Singh, B. Singh and S. Mani. 2017. Torrefaction of sorghum biomass to improve fuel properties. *Bioresour. Technol.* 232: 372–9. ISSN 0960-8524, http://dx.doi.org/10.1016/j.biortech.2017.02.060. (http://www.sciencedirect.com/science/article/pii/S0960852417301839).

Zhan, X., D. Wang, X. S. Sun, S. Kim and D. Y. C. Fung. 2002. Lactic acid production from extrusion cooked grain sorghum. *Trans. ASAE* 46:589–93.

Zhan, X., D. Wang, M. R. Tuinstra, S. Bean, P. A. Seib and X. S. Sun. 2003. Ethanol and lactic acid production as affected by sorghum genotype and location. *Ind. Crops Prod.* 18: 245–55. doi:10.1016/S0926-6690(03)00075-X.

Zhang, C., G. Xie, S. Li, L. Ge and T. He. 2010. The productive potentials of sweet sorghum ethanol in China. *Appl. Energy* 87: 2360–8. doi:10.1016/j.apenergy.2009.12.017.

Zhou, H., D. Löffler and M. Kranert. 2011. Model-based predictions of anaerobic digestion of agricultural substrates for biogas production. *Bioresour. Technol.* 102: 10819–28. doi:10.1016/j.biortech.2011.09.014.

Nutritional considerations for soybean meal use in poultry diets

Justin Fowler, University of Georgia, USA

1 Introduction

Soybeans are refined into three primary products: soybean oil, soy hulls and soybean meal (SBM). The poultry feed industry uses about 50% of the SBM produced in the United States each year. The meal is valuable to the animal feed industry because it remains an excellent source of amino acids that would otherwise be lost to the human food chain after we have removed the oil from the bean. Overall, it is estimated that (either through soybean oil or SBM) over 95% of U.S. soybean production finds its way into animal agriculture. The continued use of SBM in animal feeds depends on the ability of the meal to provide the nutritional components that make it attractive to animal producers, especially as those components are balanced with respect to the anti-nutritional factors that SBM contains. This balance has become increasingly important as price constraints have made SBM to compete with alternative ingredients such as dried distiller's grains (DDGs), as well as other oilseed meals such as canola, cottonseed or groundnut meals. In addition to these vegetable protein sources, there are a variety of protein sources such as animal by-product meals that can compete with SBM on a limited basis. Also, the amino acid requirements critical to optimal poultry performance can be met increasingly by synthetic sources in certain cases. Depending on the price considerations, it may sometimes become more advantageous to meet the essential amino acid balance of the birds with individual purified amino acids, instead of relying on SBM alone.

SBM's nutritional value is a function of both the composition of the meal and how it was processed. Despite the fact that SBM is used primarily as a source of essential amino acids in poultry feeds, there are some anti-nutritional factors that might limit its use in poultry diets. These are tied to (1) certain proteins called trypsin inhibitors that

http://dx.doi.org/10.19103/AS.2017.0034.27

are common among legumes; (2) the non-starch polysaccharides (NSP) such as mannans, galactomannans, raffinose and stachyose in the carbohydrate fraction of the meal; and (3) hormone-like substances, called phytoestrogens, that can mimic the action of the female hormone oestrogen in humans, especially with respect to egg-laying birds where the compounds can be deposited relatively efficiently in the egg (as compared to deposition in the muscle). Also, there is a growing consumer concern related to the use of genetically modified soybeans for both human and animal consumption that poses a threat to a sustainable production in the long term.

Despite the concerns that may limit the use of SBM as a feed ingredient, it still has a total digestibility by poultry of about 85% and remains the most effective single-ingredient amino acid source in poultry production. Poultry meat has been one of the most successful agricultural sectors in the last 50 years. During that time, the U.S. broiler industry has gone from small-scale, backyard businesses to a highly integrated, efficient industry which provides the most widely consumed meat in the world. Even over just the last decade, both the average market weight and feeding efficiency of broilers has continued to make significant improvements through advancements in genetic selection, rearing house management and technology, and also nutrition. The industry's demand for high-quality protein creates a competitive market for feedstuffs. Poultry companies being highly vertically integrated makes the industry sensitive to the costs of production and puts a high premium on feed ingredients of sufficient quality to keep feed conversion rates low, which is something SBM has helped achieve.

2 Nutritional content of SBM

Domestic animal agriculture uses approximately 97% of the SBM consumed in the United States each year (the poultry industry consumes about 50%). The average broiler diet contains about 25% SBM (higher levels are used in starter diets, and less is used as the birds age). The wide use of SBM as the predominant vegetable protein source in poultry diets is a function of its excellent amino acid profile and its consistency with respect to its nutrient content. The average coefficient of variation for SBM is as low as 1.7%. It is helpful for nutritionists to have an ingredient for which nutrient

Table 1 Comparative nutrient composition of soybean meal in comparison to other protein meals

	Moisture (%)	Crude protein (%)	Crude fat (%)	Dig lysine (%)	Dig methionine (%)	Dig TSAA (%)	Dig threonine (%)	AMEn (kcal/kg)
Soybean	12	47.8	1.0	3.02	0.70	1.41	2.00	2458
Cottonseed	10	41.0	2.1	1.70	0.51	1.13	1.34	2010
Canola	9	38.0	3.8	2.02	0.77	1.74	1.50	2110
Meat and bone	8	45.0	8.5	2.20	0.53	0.79	1.58	2375
Peanut	10	47.0	2.5	1.52	0.50	1.10	1.12	2677
Fish	8	62.0	9.2	4.70	1.70	2.20	2.75	2950

values do not change significantly between harvests or between seasons. Having a low coefficient of variation means that poultry producers can trust that there will not be any significant differences in the ingredient's nutrient profile. Included below is a portion of the nutrient profile of SBM from the annual Feedstuffs Ingredient Analysis Table, 2017 Edition (Table 1).

Close to half of the content of SBM is protein, and more specifically, that protein content consists of high levels of key amino acids that are important for poultry production (principally lysine). Further, not only does SBM have high amounts of these key amino acids required for poultry, the efficiency with which these amino acids are digested by the bird (the 'digestibility coefficient') are quite high. Table 2 shows the digestibility coefficients for five of the critical amino acids in poultry diets for five different oilseed meals, the ingredient with the highest value is highlighted in red.

So, not only are the total amounts of essential amino acids considered average to above average for SBM, the amino acids that it does contain are readily available for digestion when the SBM has been properly processed. The most competitive ingredient in Table 2 is canola meal, which has higher overall methionine content and, though still lower, a very similar digestibility coefficient for both methionine and lysine.

When we look at cereal grains (such as corn) alone, the bird's requirement for sulphur-containing amino acids are sufficiently met. However, the cereal grains fail to meet the bird's need for the basic amino acids, arginine and lysine. What SBM offers to poultry diets is an ingredient that has an abundance of the basic amino acids that are not provided sufficiently by staple grains such as corn. So when corn and SBM are blended at ~60:40 ratio, the only amino acids that are still deficient relative to the chicken's requirements are the sulphur-containing amino acids (methionine and cysteine) because even though the sulphur-containing amino acids in SBM are highly digestible when compared to the other oilseed meals (90% for methionine and 79% for cysteine), the total amount it contains are low relative to the requirement of the chicken, especially when compared to competitive oilseed meals such as cottonseed or canola meal.

Other plant protein sources can partially replace SBM in poultry feeds (such as cottonseed meal, peanut meal, canola meal or DDGs from corn), but these are not typically feasible unless they are used in combination with additional synthetic lysine supplementation to compensate for the reduced levels and reduced digestibility of the lysine in those alternative ingredients. However, there are anti-nutritional factors and other potential issues that may limit the use of any of these protein sources (Elkin, 2002), SBM included.

Table 2 Standardized ileal coefficients for the key amino acids in poultry. The highest coefficient in each row is highlighted in bold

	Soybean meal	Cottonseed meal	Canola meal	Meat and bone meal	Peanut meal	Fish meal
Lysine	**89**	65	80	68	76	87
Methionine	**90**	72	84	71	86	87
Cysteine	**79**	74	77	28	**79**	63
Arginine	**92**	88	87	76	91	85
Threonine	83	68	73	62	**85**	82

3 Anti-nutritive compounds

All feed ingredients contain some degree of anti-nutritive compounds that limit their use to some degree. All legumes (which include SBM) contain relatively high levels of protease inhibitors and carbohydrate-binding proteins. Fortunately, both of these are proteins and can be deactivated by heat treatment, which is why SBM will always be heat-treated before being used in poultry feeds.

The prevention of either under- or overheating SBM is critical to maintaining its desirable nutritional profile. Raw soybeans go through a process where the beans are dehulled and cracked, and the oil is extracted. The soybean oil is the primary product obtained from the beans, and this is done either through extrusion press or through solvent extraction. After the oil is extracted, the remaining meal is sent for drying and heat treatment (typically about 220°F for 90 minutes). This extra heating step is critical for keeping SBM as a desirable feed ingredient for monogastrics such as poultry, achieving a 'Goldilocks condition' of heating enough to inactive anti-nutritional compounds, but not too much so as to destroy nutrient availability.

3.1 Trypsin inhibitors

There are certain proteins called trypsin inhibitors that are common among legumes. Trypsin inhibitors can interfere with the metabolism and absorption of the protein that makes these feedstuffs useful as a source of amino acids (Liener and Kakade, 1980). Trypsin is an enzyme that the body uses to metabolize dietary proteins, thus a 'trypsin inhibitor' will prevent the proper utilization of the amino acids in the SBM by stopping the trypsin enzyme from metabolizing the dietary protein. Thus, a sufficient inactivation of trypsin inhibitors in the meal is necessary for poultry to get optimal utilization from SBM.

Luckily, trypsin inhibitors are themselves proteins, which means they can be denatured through the heat-processing step that SBM typically undergoes after the oil is extracted. If the processed meal is not heated sufficiently, then the trypsin inhibitors will not be denatured, which means they will make it into the GI tract of the bird and reduce the digestibility of the key amino acids (such as lysine) that SBM contains. The symptoms of this will be finding loose droppings that lack visible white uric acid crystals and contain visibly undigested feed.

To prevent the purchase of under-heated SBM, a urease activity index is frequently utilized by producers. The urease enzyme has no biological relationship with trypsin inhibitors, and the enzyme itself has no toxic effects to poultry. However, it is convenient that heat denatures the urease enzyme at approximately the same rate as the trypsin inhibitors, so a check for urease activity is able to provide a relatively quicker and cheaper assay that is able to indirectly determine the level of trypsin inhibitors still active in the SBM without going through the relatively more complicated assay of trypsin inhibitor activity. Thus, the lower the urease activity level (measured by a simple pH change in the sample), the less likely it is that the SBM was under-heated. A pH rise of 0.05–0.20 in the urease assay has been the accepted range for indicating sufficiently heated SBM. That lower boundary of 0.05 was initially accepted because there was a concern that if the value was lower than that, then there may be a risk of the SBM having been overheated. It was found out later on (Dale et al., 1987) that there really is no such thing as a 'lower limit' on the urease assay, as long as it is coupled with a protein solubility assay to check for the potential of the sample having been overheated (Batal et al., 2000).

The current question is if the upper boundary of 0.20 is still an acceptable indicator that birds will not be consuming sufficient trypsin inhibitors to impair their performance or if even that value may indicate too high levels of trypsin inhibitors for modern poultry diets. Based on the feed consumption of a 1950s-era broiler, and an estimation of the amount of trypsin inhibitors that would be in SBM with a 0.20 urease assay, birds at that time would have consumed about 1200 mg of trypsin inhibitors during 6 weeks. However, a modern broiler consuming SBM with a urease index at the lower limit of 0.05 will have consumed upwards of 2000 mg of trypsin inhibitors in the same amount of time. Simply put, with the current rate of feed consumption that modern broilers have, the historic range for urease activity (0.05–0.20) results in the birds consuming significantly more trypsin inhibitors than the birds that were being grown at the time the assay was first developed. Given the rate of feed consumption in modern broilers, minimizing the urease assay becomes more critical than it was 50 years ago. This makes coupling the urease assay with a measure of protein solubility all the more essential for evaluating SBM quality.

3.2 Browning reaction

Lysine is a deficient amino acid in cereal grains. It is through the use of SBM that poultry have a highly digestible source of lysine. However, the industry was cautious with respect to having an 'acceptable range' for the urease assay which evaluated trypsin inhibition of lysine digestibility because of the concern that if an accepted value was set too low, then there could be a risk of the SBM being overheated. So while all of the trypsin inhibitors will definitely have been denatured, overheating can cause its own set of issues. The ε amino group of lysine is a highly reactive part of the side chain of that amino acid and is something that is used for a variety of biological reactions unique to lysine. Crosslinks made with the ε amino group of lysine are important for forming structural proteins such as collagen or elastin, for example. However, that terminal amino group will also react with carbohydrates under excessive heat processing, which is known as the Maillard or 'browning' reaction. Once proteins undergo this reaction, they are more resistant to proteolytic digestion and, specifically, the bioavailability of lysine is significantly reduced (which was one of the amino acids that made SBM an attractive feed ingredient to begin with). The consequence is the same as what happens if there were high levels of trypsin inhibitors – all of the highly digestible lysine that made SBM a competitive feed ingredient in the first place is no longer available to the bird, and therefore makes it appear (biologically) as a poorly digestible feedstuff.

The bioavailability of lysine is therefore key to SBM's use in poultry feed. There is a possibility that some of basic amino acid analyses can be inaccurate in estimating the lysine content of heat-modified feedstuffs. Only lysine with a free ε amino group is considered biologically available for protein synthesis (Rutherford and Moughan, 2007; Kim et al., 2012). Rutherfurd (2015) evaluated the importance of determining both reactive lysine and bioavailable lysine and found that the guanidination reaction is the most suitable for determining the lysine content that is bioavailable. The guanidination reaction targets the unreacted ε amino group of lysine through a reaction of O-methylisourea with lysine to produce homoarginine. It has been found to be another basis for determining unreacted versus bioavailable lysine.

In total, the safest practice for a producer to evaluate SBM quality is to couple the urease assay (where a high value will indicate an under-heated SBM) with a measure of protein solubility or availability (where a low value indicates an overheated SBM).

3.3 Non-starch polysaccharides

'Fibre' is simply a shorthand way of referring to the carbohydrate fraction of a feed ingredient that is not readily digestible by the animal via endogenous means. Dietary fibre is made up of a diverse mixture of carbohydrate polymers that may or may not be associated with non-carbohydrate groups (such as proteins or fatty acids), which are typical in plant cell walls. As a group, this set of carbohydrates are referred to by the collective term of NSP. Though we think of SBM as a protein source, and think of NSP as a problem related to the more carbohydrate-rich ingredients such as corn or wheat, the carbohydrate fraction of SBM can affect its use as well. SBM contains relatively high levels of NSP, approximately 20% (3% soluble NSP and 16% insoluble NSP). These NSP consist mainly of mannans and galactomannans, as well as two indigestible galactosides: raffinose and stachyose (typically present at about 1% and 5% of the SBM, respectively).

Since poultry have a relatively high rate of passage for digestive material (approximately 3–4 hours) and lack a significant ability to utilize microbial assistance in the way that ruminants, horses or ratites can, the presence of NSP in the diet can result in reductions in their nutrient utilization and performance. The harm NSP can cause, biologically, is related to the increase in the intestinal viscosity that occurs when they are included at high levels in the diet. When the material in the digestive tract is more viscous: enzyme contact with food components is decreased, overall nutrient absorption is reduced and the turnover rate of intestinal cells is higher (Gee, 1996; Daenicke, 2000; Teirlynck, 2009), and the absorption of any of the other nutrients in the diet is reduced. When these effects of viscosity combine, it has been shown that a 4% increase in NSP content can result in a 10% reduction in the digestibility of the lipid, starch and protein components of a feed. Maximizing the nutrient utilization from ingredients in the diet is a necessity when it comes to feeding poultry. Doing so not only maximizes the productive efficiency of the birds (which is important given that feed cost can be as much as 70% of the cost of production) but also will reduce nutrient excretion into the litter (which can have negative environmental impacts once that litter is land-applied).

One option available to nutritionists to make the most of the NSP content in feed ingredients is to use exogenous NSP-degrading enzymes (NSP-ase). NSP-ase function by hydrolysing the carbohydrate polymers in the plant cell wall into smaller carbohydrate fragments, which creates an impact on the increase in viscosity that is characteristic of NSP. The criteria that determine a 'good' NSP-ase for poultry can be quite limiting. For one, they must be able to pass through the low pH of the upper GI tract and the high temperatures of feed formulation, and they also must work during the limited interaction time with the NSP substrates allowed for by the high rate of passage of digestive material in the bird's GI tract. So, among the thousands of enzymes that have been discovered, it is a much smaller number that are true candidates for inclusion in poultry diets (Leeson and Summers, 2001).

The enzymes that have potential for exogenous use in poultry diets have a set of key methods of action: hydrolysing chemical bonds that are not sufficiently degraded by endogenous mechanisms, eliminating the encapsulation of key nutrients by undigested compounds, breaking down insoluble NSP for later digestion in the hindgut and supporting the activity of endogenous enzymes that may have not yet reached full activity in younger animals. Latham et al. (2016) found that improvements in broiler growth performance with the inclusion of an NSP-ase (β-mannanase) were most pronounced in the earlier phases of production where the levels of SBM are at their highest. Further a review by Slominski

(2011) found that an NSP-ase 'cocktail' of enzymes can be more effective at improving the performance of birds fed even a highly digestible, corn-SBM diet.

Dietary NSP are non-digestible by poultry, and though the concentrations are relatively low in corn-SBM diets, it has been shown that a combination of enzymes improves broiler performance and energy digestibility (Kaczmarek et al., 2014). However, those authors noted that differences in particle size can contribute to differences in the effects reported in the literature. Though SBM is most often considered as a primary source of protein, its carbohydrate fraction contains close to 20% NSP (Irish and Balnave, 1993). Furthermore, De Keyser et al. (2016) found that using two-fold the recommended dosage of an NSP-ase resulted in an increased efficiency of total protein and amino acid digestibility, suggesting that liberating non-digestible carbohydrates can also improve protein utilization.

3.4 Isoflavones

Isoflavones are phytoestrogens (plant-derived molecules with oestrogen-mimicking activity) found primarily in legumes. Among the legumes, soybeans have the highest isoflavone concentration (ranging from 0.12% to 0.31%). There are predominantly twelve isoflavone compounds in soybeans, and those that have been particularly studied are glyciein, genistein, daidzein and a highly active daidzein-metabolite called equol. Their consumption has been linked to both beneficial effects in the treatment of chronic diseases such as cancer, osteoporosis and cardiovascular disease (Anderson and Garner, 1997; Adlercreutz, 1998; Potter et al., 1998), as well as negative effects on reproductive functions in animal studies (Davies and Hill, 1989; Urpi-Sarda et al., 2008) due to its interaction with oestrogen receptors.

While the health impacts of soy isoflavones are still open to investigation, there is undeniable pressure from consumers with regard to the use of soybeans. Consumers are becoming more selective in their diet choices, and there is an emerging concern among a segment of the market with respect to the presence of soy isoflavones in the diet. Studies that have examined the effects of these soy-derived phytoestrogens on human health have been inconclusive, and soybeans are not the only sources of phytoestrogen in the human diet (other sources include coffee, tea, nuts, flax, sesame, sunflower, garlic, onion, red wine, cabbage, broccoli, berries and alfalfa). What is unique to poultry is the fact that the laying hen is capable of efficiently packaging the contents of her diet into the eggs she produces. This is a fact that we have utilized to increase the content of beneficial nutrients in the human diet (such as omega-3 fatty acids and fat-soluble vitamins), as well as to alter the yolk colour to meet consumer preferences. This means that it is also possible for soy isoflavones remaining in SBM to be transferred from the diet and into the egg. Whether consumers consider these compounds beneficial or not, the egg may represent a source of isoflavones in a typical person's diet.

4 Genetically modified soybeans

A genetically modified organism (GMO) is defined as any organism that has had its genome altered by genetic engineering techniques. In its most basic form, genetic modification of plants and animals has been happening for thousands of years as a part of human agriculture, modifying the ancestral teosinte into the modern corn, for example. But, more specifically, 'GMO' tends to most often refer to plants that have had genes from other species inserted into their genome (rather than evolution through selective breeding). Genetic modification has significant implications for the sustainability of agriculture. It has been predicted for quite some time that food production will need to double by 2025 and triple by 2050 (Vasil, 1998). Both grain and oilseed production has shown increased yields in response to GM interventions, using less inputs to produce more output.

With respect to soybeans, Monsanto introduced the first GM soybean in the United States in 1994, and within 20 years, close to 90% of the total soybean cultivation utilized GM soybeans. 'Roundup Ready Soybeans' were produced to be resistant to the glyphosate herbicide, which interferes with the synthesis of aromatic amino acids. These amino acids are essential in animals (must be consumed in the diet), but can be produced naturally by plants, and glyphosate interferes with that production. Also, Edwards et al. (2000) evaluated a SBM from a GM line that was designed to result in higher crude protein.

Though there is consensus that food derived from GMOs pose no negative health-related impact to humans (Nicolia, et al., 2014), the average consumer is much less likely to view GMOs as 'safe', which has led to regulatory and labelling standards that differ greatly from country to country. During the process of digestion, proteins are broken down into tri- and dipeptides, plus free amino acids. So any proteins arising from the genetic modification should be completely destroyed in the upper GI tract. Ash et al. (2003) found no GM proteins from SBM in whole egg, egg albumin, liver samples or faeces of laying hens fed GM soybeans (even though the GM proteins could be found in the feed).

5 Conclusion and future trends

The poultry industry provides a significant portion of the global demand for soybeans, thus the sustainability of soybean production (as a whole) is linked to the value of SBM as a feed ingredient in animal agriculture. The excellent benefits that SBM brings as a protein source can only be capitalized on when it is balanced with the anti-nutritional factors that it contains, which is a function of how it was processed. These limiting factors include trypsin inhibitors, NSP and concerns over phytoestrogens and genetic modification.

The poultry industry's demand for high-quality protein creates a competitive market for feedstuffs. Despite the concerns that may challenge the use of SBM as a feed ingredient, it still has a total digestibility by poultry of about 85% and remains the most effective single-ingredient amino acid source in poultry production. The use of exogenous enzymes has also served to improve the competitiveness of SBM in poultry diets by shifting the carbohydrate fraction away from non-digestible NSP.

Future research topics should focus on what would maintain SBM's competitiveness within the animal feed market. Particularly, given the characteristics that make SBM beneficial for feeding poultry, attention should be given to those features as they relate to the potential for alternative ingredients to become more competitive in feed formulation.

Also, more research can be done with respect to the effects of trypsin inhibitors in fast-growing genetic lines with much higher rates of feed consumption than 20 or 30 years ago when initial estimations of acceptable values for the urease assay were first described.

6 Where to look for further information

There are organizations presently involved in the development of strategies to maintain a stable supply of soybeans for both human and animal agriculture needs: The United Soybean Board, USAID's Soybean Innovation Lab, U.S. Farmers and Ranchers Alliance, and the U.S. Soybean Export Council.

7 References

Adlercreutz, H. 1998. Epidemiology of phytoestrogens. *Bailliers Clin. Endocrinol. Mateb.* 12:605–23.

Anderson, J., and S. Garner. 1997. Phytoestrogens and human function. *Nutr. Today* 32:232–9.

Ash, J., C. Novak and S. E. Scheideler. 2003. The fate of genetically modified protein from roundup ready soybeans in laying hens. *J. Appl. Poult. Res.* 12:242–5.

Batal, A. B., M. W. Doublas, A. E. Engram and C. M. Parsons. 2000. Protein dispersibility index as an indicator of adequately processed soybean meal. *Poult. Sci.* 79:1592–6.

Daenicke, S., W. Bottcher and H. Jeroch. 2000. Replacement of soybean oil with tallow in rye-based diets without xylanase increases protein synthesis in small intestine of broilers. *J. Nutr.* 130:827–34.

Dale, N., M. Araba and E. Whittle. 1987. Protein solubility as an indicator of optimum processing of soybean meal. *Proceedings of the 1987 Georgia Nutrition Conference for the Feed Industry.* Atlanta, GA, pp. 88–95.

Davies, H. and J. Hill. 1989. The effect of diet on the metabolism in sheep of the tritiated isoflavones formononetin and biochanin A. *Aust. J. Agric. Res.* 40:157–63.

De Keyser, K., L. Kuterna, S. Kaczmarek, A. Rutkowski and E. Vanderbeke. 2016. High dosing NSP enzymes for total protein and digestible amino acid reformulation in a wheat/corn/soybean meal diet in broilers. *J. Appl. Poult. Res.* 25:239–46.

Edwards, H. M., M. W. Douglas, C. M. Parsons and D. H. Baker. 2000. Protein and energy evaluation of soybean meals processed from genetically modified high-protein soybeans. *Poult. Sci.* 79:525–7.

Elkin, R. G. 2002. Nutritional components of feedstuffs: a qualitative chemical appraisal of protein. In *Poultry Feedstuffs Supply, Composition and Nutritive Value.* Eds McNab, J. and Boorman, N. Poultry Science Symposium Series, volume 26.

Gee, J. M., W. Lee-Finglas and G. W. Wortley. 1996. Fermentable carbohydrates elevate plasma enteroglucagon but high viscosity is also necessary to stimulate small bowel mucosal cell proliferation in rats. *J. Nutr.* 126:373–9.

Irish, G. G., and D. Balnzve. 1993. Non-starch polysaccharies and broiler performance on diets containing soybean meal as the sole protein concentrate. *Aust. J. Agric. Res.* 44:1483–99.

Kaczmarek, S. A., A. Rogiewicz, M. Mogielnicka, A. Rutkowski, R. O. Jones and B. A. Slominski. 2014. The effect of protease, amylase, and nonstarch polysaccharides-degrading enzyme supplementation on nutrient utilization and growth performance of broiler chickens fed corn-soybean meal-based diets. *Poult. Sci.* 93:1745–53.

Kim, J. C., B. P. Mullan and J. R. Pluske. 2012. Prediction of apparent, standardized, and true ileal digestible total and reactive lysine contents in heat-damaged soybean meal samples. *J. Anim. Sci.* 90:137–9.

Latham, R. E., M. Williams, K. Smith, K. Stringfellow, S. Clemente and R. Brister, J. T. Lee. 2016. Effect of β-mannanase inclusion on growth performance, ileal digestible energy, and intestinal viscosity of male broilers fed a reduced-energy diet. *J. Appl. Poult. Res.* 25:40–7.

Leeson, S., and J. D. Summers. 2001. *Scott's Nutrition of the Chicken.* 4th edn. Guelph, ON: University of Guelph.

Liener, I. E., and Kakade M. L. 1980. Protease inhibitors. In *Toxic Constituents of Plant Foodstuffs.* Ed. Liener, I. E. New York, USA: Academic Press.

Nicolia, A., A. Manzo, F. Veronesi and D. Rosellini. 2014. An overview of the last 10 years of genetically engineered crop safety research. *Crit. Rev. Biotechnol.* 34:77–88.

Potter, S., J. Baum, H. Teng, R. Stillman, N. Shay and J. Erdman. 1998. Soy protein and isoflavones: their effects on blood lipids and bone density in postmenopausal women. *Am. J. Clin. Nutr.* 68:1375S–9S.

Rutherford, S., and P. J. Moughan. 2007. Development of a novel bioassay for determining the available lysine contents of foods and feedstuffs. *Nutr. Res. Rev.* 20:3–16.

Rutherford, S. 2015. Use of the guanidination reaction for determining reactive lysine, bioavailable lysine and gut endogenous lysine. *Amino Acids.* 47:1805–15.

Slominski, B. A. 2011. Recent advances in research on enzymes in poultry diets. *Poult. Sci.* 90:2013–23.

Teirlynck, E., L. Bjerrum and V. Eeckhaut. 2009. The cereal type in feed influences gut wall morphology and intestinalimmune cell infiltration in broiler chickens. *Br. J. Nutr.* 102:1453–61.

Urpi-Sarda, M.; C. Morand, C. Besson, G. Kraft, D. Viala, A. Scalbert, J. Besle and C. Manach. 2008. Tissue distribution of isoflavones in ewes after consumption of red clover silage. *Arch. Biochem. Biophys.* 476:205–10.

Vasil, I. K. 1998. Biotechnology and food security for the 21st century: A real-world perspective. *Nat. Biotechnol.* 16:399–400.

www.ingramcontent.com/pod-product-compliance
Lightning Source LLC
Chambersburg PA
CBHW050536270326
41926CB00015B/3249